Algebraic Number Theory

Textbooks in Mathematics

Series editors:
Al Boggess, Kenneth H. Rosen

https://www.routledge.com/Textbooks-in-Mathematics/book-series/CANDHTEXBOOMTH

Algebraic Number Theory
A Brief Introduction

J. S. Chahal

CRC Press
Taylor & Francis Group
Boca Raton London New York

CRC Press is an imprint of the
Taylor & Francis Group, an **informa** business

A CHAPMAN & HALL BOOK

First edition published 2022
by CRC Press
6000 Broken Sound Parkway NW, Suite 300, Boca Raton, FL 33487-2742

and by CRC Press
2 Park Square, Milton Park, Abingdon, Oxon, OX14 4RN

© 2022 Taylor & Francis Group, LLC

CRC Press is an imprint of Taylor & Francis Group, LLC

ISBN: 978-0-367-76145-5 (pbk)
ISBN: 978-1-003-17703-6 (ebk)

Typeset in CMR10 font
by KnowledgeWorks Global Ltd.

Contents

Preface

During the last decade of the last millennium, I taught a course on algebraic number theory, first at the University of Salzburg and the following year at Brigham Young University, for which I had prepared my own notes. They were inspired by two famous classics: (1) Alan Baker, *Transcendental Number Theory* and (2) Atiyah-MacDonald, *Commutative Algebra*. Both have covered the essentials of the subjects in as few pages as possible. In 2003, my notes were published by Kendrick Press (a local publisher in Utah striving to make available rare books in mathematics, e.g., the first English translation of Riemann's collected papers). This book is an expanded version of these notes. Thanks to the referees at whose suggestions now the first two chapters have been added for the convenience of the reader. Also, at their suggestion, the last chapter has been expanded considerably. I hope the book will be helpful at least to those who do not have time to plough through voluminous treatises but would like to know the basics of algebraic number theory.

We have tried to keep the treatment of the subject (Chapters 3–9) as classical as possible. In this form, it was developed originally by the German school, as summed up in Hilbert's *Zahlbericht*. (See [24] in the Bibliography.) Later, the theory was recast, from a different and more abstract point of view. The newer approach, for which Weil's *Basic Number Theory* [39] may be consulted, provides a broader theory to include the arithmetic of algebraic curves over finite fields, and even suggests a theory for studying higher dimensional varieties over finite fields. However, if one is not familiar with the classical algebraic number theory (of number fields), Weil's book may be difficult to read. Commutative algebra also originated in algebraic number theory. The purpose of the last chapter is to indicate how the subject treated in this book leads naturally to the Weil conjectures and some delicate questions in algebraic geometry. We shall discuss, without supplying complete details, some advantages of this approach to algebraic number theory.

This book is suitable for an independent study or as a textbook for a small class. We present the subject as developed by its creators Dedekind, Dirichlet, Hilbert, Kronecker, Minkowski and Weber, etc. Besides a basic knowledge of elementary number theory and linear algebra, a first course in abstract algebra and in Galois theory (e.g., [40], [9], [23], and [1] or [31]) should more than suffice as a prerequisite.

I would like to thank Professor Darrin Doud for reading the final version of my notes published in 2003 and pointing out several misprints, and Professor Roger Baker for doing an outstanding job of proofreading. Thanks to Professor Wolfgang Herfort for thoroughly reading the current version of the book to point out errors and suggest improvements. I would also express my gratitude to Professor Fritz Schweiger for inviting me to the University of Salzburg. Last but not least, many thanks to Lonette Stoddard for her outstanding job with LaTeX.

J. S. Chahal
Provo, UTAH
21 December 2020

1

Genesis: What Is Number Theory?

1.1 What Is Number Theory?

Number Theory is the study of numbers, in particular the *whole numbers* $1, 2, 3, \ldots$, also called the *natural numbers*. The set of natural numbers is denoted by $\mathbb{N}$. Leaving aside the unit 1, these numbers fall into two categories: The indivisible numbers $2, 3, 5, 7, \ldots$ are the *primes*, and the rest $4, 6, 8, 9, 10, \ldots$ composed of primes, are the *composite numbers*. The following basic facts, with proofs, about these numbers were already known to Euclid around 300 B.C.

Theorem 1.1. *There are infinitely many primes.*

Theorem 1.2 (Fundamental Theorem of Arithmetic). *Every natural number $n > 1$ is a unique product*

$$n = p_1^{e_1} \ldots p_r^{e_r} \quad (r \geq 1) \tag{1.1}$$

of powers of distinct primes $p_1, \ldots, p_r$, taken in some order.

By looking at the list of primes, one can ask several naive but still unanswered questions. For example, is there an endless supply of twin primes? We call a pair of primes q, p *twin primes* if $p = q + 2$. [This is the closest two odd primes can be to each other.] A glance at the list

$$3, 5; \quad 5, 7; \quad 11, 13; \quad 17, 19; \quad 29, 31; \ldots$$

suggests that there are infinitely many pairs of twin primes, but no one has ever been able to prove this so far. Another big problem in number theory is the unproven conjecture of Goldbach, which asserts that every even number larger than 2 is a sum of two primes.

Many questions in number theory arise naturally in the study of geometry. The most fundamental fact in Euclidean geometry is the theorem of Pythagoras, which may be called the fundamental theorem of geometry. Actually, it was known to the Egyptians and Babylonians about two thousand years earlier, but they had no rigorous proof of it like Euclid did.

1

Theorem 1.3 (Fundamental Theorem of Geometry). *The real numbers* $0 < x \le y < z$ *are the side lengths of a right triangle if and only if*

$$x^2 + y^2 = z^2. \tag{1.2}$$

To number theorists, the most interesting solutions of (1.2), called the *Pythagorean triples*, are those with x, y, z whole numbers, such as $(3, 4, 5)$, $(5, 12, 13)$. It is said that the Egyptians used long ropes divided into three parts by two knots of lengths 3, 4 and 5 units. They knew that this way they could get error-free right angles for marking the bases for their huge pyramids.

If (x, y, z) is a Pythagorean triplet, so is (cx, cy, cz) for all integers $c > 0$. But up to similarity, they all represent the same right triangle. The smallest among them is called a *primitive Pythagorean triplet*. One can ask if there are infinitely many primitive Pythagorean triplets. The Babylonians were aware of at least 15 of them. Since it is exceedingly difficult to find them by trial and error, they must have had an algorithm to produce them. Diophantus of Alexandria (third century A.D.) for sure had the algorithm to generate them all [cf. [41]].

By the squares and the cubes we shall mean the squares and cubes of whole numbers. Thus the above statement about Pythagorean triples is a statement about splitting a square into a sum of two squares. Mathematics of the Islamic world during its golden age was built upon the work of the Greek (Euclid, Archimedes, Diophantus) and Indian (Aryabhata and Brahmagupta) mathematicians. According to Dickson [13], the Islamic mathematician al-Khujandi (from Khujand, Tajikistan) was the first to claim that a cube cannot be split into two cubes, and gave an erroneous proof. However, it was Fermat who claimed that for all integer exponents $n \ge 3$, the equation

$$x^n + y^n = z^n \tag{1.3}$$

has no nontrivial solutions in the set $\mathbb{Z} = \{0, \pm 1, \pm 2, \ldots\}$ of *integers*. This came to be known as Fermat's Last Theorem or FLT for short. For $n = 4$, Fermat gave a proof by showing that even the equation

$$x^4 + y^4 = z^2 \tag{1.4}$$

has no nontrivial solutions in integers. The proof is based on the Fundamental Theorem of Arithmetic (cf. Exercise 3). It is easy to see that it suffices to prove FLT when the exponent $n = 4$ or n is a prime > 3.

It was realized by Euler, Gauss and others that to prove FLT for $\ell = 3$, the field $\mathbb{Q}$ of rational numbers is not adequate and one must deal with complex numbers and wish for the unique factorization in rings like $\mathbb{Z}[\sqrt{-3}] = \{a + b\sqrt{-3} \mid a, b \in \mathbb{Z}\}$. Unfortunately, this is not true: $4 = 2 \cdot 2 = (1 + \sqrt{-3})(1 - \sqrt{-3})$ has two factorizations in $\mathbb{Z}[\sqrt{-3}]$.

To remedy this, one sees that the field $K = \mathbb{Q}(\sqrt{-3}) = \{r+s\sqrt{-3} \mid r, s \in \mathbb{Q}\}$ is a field of fraction of $\mathbb{Z}[\sqrt{-3}]$ but also of a slightly bigger ring B, which is a unique factorization domain (UFD), of the so-called *Eisenstein integers* $\mathbb{Z}[\omega] = \{a + b\omega \mid a, b, \in \mathbb{Z}\}$, where $\omega = -\frac{1}{2} + \frac{\sqrt{3}}{2}i$ is a primitive cube root of unity, i.e. $\omega \neq 1$ is a root of the polynomial $u^3 - 1$.

The idea of Euler to prove FLT for the exponent $\ell = 3$ was to factor the left-hand side of $x^3 + y^3 = z^3$ as $x^3 + y^3 = (x + y)(x + \omega y)(x + \omega^2 y)$ and then use the divisibility arguments in the ring $\mathbb{Z}[\omega]$ that Fermat used in his proof for the exponent $n = 4$ in $\mathbb{Z}$.

The following characterization of the ring B above was crucial in the development of algebraic number theory.

Theorem 1.4. *The ring $\mathbb{Z}[\omega]$ is the set of all algebraic integers in the field $K = \mathbb{Q}(\sqrt{-3})$, i.e. the elements α of K that satisfy monic polynomials over $\mathbb{Z}$.*

A polynomial is *monic* if its leading coefficient is 1. Clearly ω is an algebraic integer, but it is not in $\mathbb{Z}[\sqrt{-3}]$.

Remark 1.5. By this definition, the set of algebraic integers in $\mathbb{Q}$ is the ring $\mathbb{Z}$, as desired.

One can try to prove FLT for all odd prime exponents by factoring its left-hand side as

$$x^\ell + y^\ell = (x + y)(x + \zeta y) \cdots (x + \zeta^{\ell-1}y) \tag{1.5}$$

in the ring $\mathbb{Z}[\zeta] = \{a_0 + a_1\zeta + \cdots + a_{\ell-1}\zeta^{\ell-1} \mid a_j \in \mathbb{Z}\}$, where $\zeta = \zeta_\ell = \cos\frac{2\pi}{\ell} + i\sin\frac{2\pi}{\ell}$ is a primitive ℓ-th root of 1. However, it turned out that the unique factorization does not hold in general for all $\mathbb{Z}[\zeta_\ell]$. In fact, it holds if and only if $\ell \leq 19$. There is a larger class of primes, which Kummer called regular primes and proved FLT for. This was quite an achievement because only three primes 37, 59 and 67 under 100 are irregular. Although there is an algorithm to check whether or not a given prime ℓ is regular [cf. Exercise 5], it is not known if there are infinitely many regular primes, whereas the infinitude of irregular primes has been known for some time.

Rather than trying to prove FLT, the school of nineteenth century German mathematicians tried to build a theory of rings (called the Dedekind domains) to recover the unique factorization for these rings. For this, Dedekind replaced the integers in $\mathbb{Z}$ by ideals $\mathfrak{a}$ and the primes $p = 2, 3, 5, \ldots$ by nonzero prime ideals $\mathfrak{p}$ in these rings. He then showed that every nonzero and non-unit ideal has a unique factorization

$$\mathfrak{a} = \mathfrak{p}_1^{e_1} \ldots \mathfrak{p}_r^{e_r}$$

as powers of distinct prime ideals $\mathfrak{p}_1, \ldots, \mathfrak{p}_r$, taken in some order.

In the ring $\mathbb{Z}$, the integers $a \geq 0$ and primes $p = 2, 3, 5, \ldots$ can be identified by the principal ideals $\mathfrak{a} = (a) = a\mathbb{Z} = \{am \mid m \in \mathbb{Z}\}$ and the prime ideals $\mathfrak{p} = (p) = p\mathbb{Z}$. However, the ideal (p) generated by a rational prime $p = 2, 3, 5, \ldots$ may not be a prime ideal in the ring $\mathcal{O}$ of integers of a number field K. A *number field* by definition is a finite extension of the field $\mathbb{Q}$ of rational numbers. So the following question needed an answer: What is the unique factorization

$$(p) = \mathfrak{p}_1^{e_1} \ldots \mathfrak{p}_r^{e_r}$$

of (p) into a unique product of power of prime ideals $\mathfrak{p}_1, \ldots, \mathfrak{p}_r$ in the ring $\mathcal{O}$? [The Germans called this ring an *order* and denoted it by $\mathcal{O}$.] At the end of the next chapter we will discuss the ring $\mathbb{Z}[i]$, $i = \sqrt{-1}$, of Gaussian integers for clues to answer this question.

A *domain* (commutative ring with 1 but without zero divisors) A is a *principal ideal domain* (or PID for short) if every ideal of A is generated by one element, i.e. is of the form $(a) = \{ax \mid x \in A\}$. A domain is a *unique factorization domain* (UFD) if the unique factorization of nonzero, non-unit elements into a unique product of powers of distinct primes holds. It is well known that a PID is a UFD (but not conversely). So the Germans were interested in knowing how far the ring $\mathcal{O}_K$ of integers of a number field K can be from being a PID. They associated to each number field K a positive integer h_K, called the class number of K, which measures the deviation of $\mathcal{O}_K$ from being a PID. In particular, $\mathcal{O}_K$ is a PID if and only if $h_K = 1$. They proved that h_K is always finite. Kummer called an odd prime ℓ *regular* if it is not a factor of the class number of the field $\mathbb{Q}(\zeta_\ell) = \{c_0 + c_1\zeta_\ell + \cdots + c_{\ell-1}\zeta_\ell^{\ell-1} \mid c_j \text{ in } \mathbb{Q}\}$ and proved FLT for all regular primes. (For proof, see [5, pp. 156 & 378].)

1.2   Methods of Proving Theorems in Number Theory

1. Unique Factorization Arguments

The method that has been used since antiquity is the unique factorization. Let us recall Euclid's proof of Theorem 1.1.

It follows from the unique factorization (1.1) that any $n > 1$ is either a prime or has a prime factor. To prove Theorem 1.1 by contradiction, suppose there are only finitely many primes, say $p_1, \ldots, p_r$. Now consider the number $n = p_1 \ldots p_r + 1$. It is not a prime because it is larger than every prime p_j. So, it has a prime factor, say p_1. Therefore $n = p_1 a$ for an integer a. This implies that $1 = p(a - p_2 \ldots p_r)$. This is a contradiction because 1 has no prime factor. ∎

Another example of such a proof is the proof below by Euler (1770) of the following claim of Fermat (1657): 27 is the only cube that exceeds a square by 2. In modern terminology, $(3, \pm 5)$ are the only points with integer coordinates on the elliptic curve

$$y^2 = x^3 - 2. \tag{1.6}$$

Proof. In the ring $\mathbb{Z}[\sqrt{-2}] = \{a + b\sqrt{-2} \mid a, b \in \mathbb{Z}\}$, which is a UFD (see Exercise 8, Chapter 2), we use the factorization

$$x^3 = y^2 + 2 = (y + \sqrt{-2})(y - \sqrt{-2}).$$

In general, in a UFD, if α, β have no common factor other than units, and $\alpha\beta = \gamma^m$ for an integer $m > 0$, then $\alpha = \alpha_1^m$ and $\beta = \beta_1^m$ for some α_1, β_1 in it. Therefore

$$y + \sqrt{-2} = (a + b\sqrt{-2})^3 \text{ for } a, b \in \mathbb{Z}.$$

By expanding $(a + b\sqrt{-2})^3$ and comparing the real/imaginary parts, we get

$$1 = b(3a^2 - 2b^2), \quad y = a^3 - 6ab^2. \tag{1.7}$$

But the first equation in (1.7) can hold only if $b = 1$ and $a = \pm 1$. This implies $y = \pm 5$. $\qquad\square$

2. Analytic Methods

Euler initiated what we call the analytic number theory. The study of infinite series (analysis) can lead to interesting results in number theory. Let us recall Euler's proof of the infinitude of primes. Leaving aside the issue of convergence, by multiplying the infinite series formally, one sees that

$$\sum_{n=1}^{\infty} \frac{1}{n} = \sum \frac{1}{p_1^{e_1} \cdots p_r^{e_r}} = \prod_p \left(1 + \frac{1}{p} + \frac{1}{p^2} + \cdots\right), \text{ i.e.}$$

$$\sum_{n=1}^{\infty} \frac{1}{n} = \prod_p \left(1 - \frac{1}{p}\right)^{-1}, \tag{1.8}$$

the product (called the Euler product) taken over all primes p. Note that the first equality is a consequence of the unique factorization (1.1).

The partial sums $\sum_{n=1}^{N-1} \frac{1}{n}$ of the series $\sum_{n=1}^{\infty} \frac{1}{n}$ are bounded from below by the area (cf. Figure 1.1) $\int_1^N \frac{dx}{x} = \ln N$, which goes to infinity as N goes to infinity.

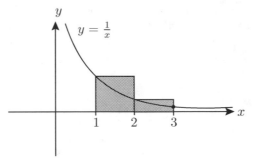

FIGURE 1.1: Divergence of $\sum\limits_{n=1}^{\infty} \frac{1}{n}$.

Now if there were only finitely many primes, the right-hand side of (1.8) is finite whereas its left-hand side is infinite. This is a contradiction.   ∎

Let $\phi(m)$ be the Euler ϕ-function. For a positive integer m, it is the number of integers a $(1 \le a < m)$ such that a and m have no common factor > 1. For such an a, consider the semi-residue class $S_a = \{a + bm \mid b \in \mathbb{N}\} \bmod m$. In 1857 Dirichlet used the Dirichlet L-series to prove that each S_a contains infinitely many primes. Moreover, each S_a contains its expected share $\frac{1}{\phi(m)}$ of primes.

3. Techniques from Algebraic Geometry

Algebraic geometry is the study of the solutions of polynomial equations in a number of variables $x_1, \ldots, x_n$ with values of x_j in a field K. Unless we assume K to be algebraically closed, such as the field $\mathbb{C}$ of complex numbers, the subject is not satisfactory. For example, $x^2 + y^2 + 1 = 0$ has no solution with x, y even in such a big field as $\mathbb{R}$, the field of real numbers. Moreover, a line (equation of degree 1) is supposed to meet a circle (equation of degree 2) in two points. This rarely happens, but happens every time (in the projective plane $\mathbb{P}^2(\mathbb{C})$), thanks to *Bezout's Theorem: Two curves of degree d_1, d_2 with no component in common intersect in $d_1 d_2$ points in the projective plane $\mathbb{P}^2(\mathbb{C})$, counted properly.*

The arithmetic algebraic geometry is the subject in which algebraic geometric methods are used to answer questions in number theory. We illustrate it by finding the primitive Pythagorean triples, which is the same as finding the rational points (points with rational coordinates) on the unit circle

$$X^2 + Y^2 = 1, \tag{1.9}$$

with the rational numbers X, Y in the lowest form. A primitive Pythagorean triple (x, y, z) gives such a rational point with $X = \frac{x}{z}$, $Y = \frac{y}{z}$, and vice versa.

To obtain an algorithm to find all the primitive Pythagorean triples (x, y, z), we parameterize the unit circle (1.9) by the slope t of the line through the fixed point $(-1, 0)$ and a variable point (X, Y) on this circle (cf. Figure 1.2).

Substituting for X from the equation $X = tY - 1$ of this line in equation (1.9) of the unit circle, an easy calculation shows that

$$Y = \frac{2t}{1 + t^2} \quad \text{and} \quad X = tY - 1 = \frac{1 - t^2}{1 + t^2}.$$

If we run t through all rational numbers in the lowest form $t = \frac{a}{b}$, we get the following result:

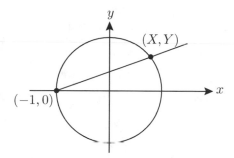

FIGURE 1.2: Rational points on the unit circle.

Theorem 1.6. *Every primitive Pythagorean triplet (x, y, z) is of the form*

$$x = a^2 - b^2, \ y = 2ab, \ z = a^2 + b^2,$$

where a, b $(a > b)$ are positive integers of opposite parity (one odd, the other even) with no common factor.

Note that the condition of opposite parity is necessary because otherwise x, y, z are all even, so (x, y, z) is not primitive. We also remark that switching x and y does not produce a different Pythagorean triplet.

EXERCISES

1. Verify equation (1.5).

2. Use Theorem 1.6 to find 15 primitive Pythagorean triplets (the same number of them as the Babylonians did).

3. **Proof of FLT for the exponent $n = 4$**

   Example of Fermat's method of descent:

   It is enough to show that $x^4 + y^4 = w^2$ has no solution either with $x, y, w > 0$. If it does, pick one with the smallest w.

i) Show that x^2, y^2 and w is a primitive Pythagorean triplet, hence by Theorem 1.6,

$$x^2 = a^2 - b^2, \ y^2 = 2ab, \ w = a^2 + b^2$$

with $a > b > 0$ coprime and of opposite parity.

ii) Show that Theorem 1.6 is applicable again to $x^2 + b^2 = a^2$ to get

$$x = s^2 - t^2, \ b = 2st, \ a = s^2 + t^2$$

with $s > t > 0$ coprime and of opposite parity.

iii) Show that s, t and a are relatively coprime, hence $y^2 = 4ast$ implies that s, t and a are all squares, say u^2, v^2 and c^2.

Hence, $u^4 + v^4 = c^2$, which contradicts the minimality of w.

4. (a) Show that $y^2 = x^3 - 5$ has no solution in integers.

  (b) Show that the only integer points on the elliptic curve

$$y^2 = x^3 - 4$$

  are $(5, \pm 11)$ and $(2, \pm 2)$.

  Hint: Factor

$$y^2 + 4 = (2 + iy)(2 - iy)$$

  in $\mathbb{Z}[i]$ and consider the cases of y odd and even.

5. The rational numbers $B_m (m \geq 1)$, defined by

$$\frac{e^x}{e^x - 1} = 1 + \sum_{m=1}^{\infty} \frac{B_m}{m!} x^m$$

are called the *Bernoulli numbers*. An odd prime ℓ is a *regular prime* $\Leftrightarrow \ell$ does not divide the numerators of the Bernoulli numbers $B_2, B_4, \ldots, B_{\ell-3}$ (taken in the lowest form). Use this criterion to show that 31 is a regular prime, but 37 is not.

2

Review of the Prerequisite Material

The reader is expected to be familiar with elementary number theory (cf. [40]), the language of groups, rings, fields, [23] and vector spaces as found e.g. in [9].

However, to avoid later interruptions we briefly review these concepts, mostly without proofs. Our main purposes in doing so are to i) make clear what is meant by certain mathematical terms, and ii) state the theorems in the form needed for their applications.

2.1 Basic Concepts

A *group* is a pair $(G, *)$ of a nonempty set G and a *binary operation* $*$ on G, i.e. a map $G \times G \ni (x, y) \to x * y \in G$, called the *group law* on G with the following properties:

i) The group law is *associative*: for all x y, z in G, $(x * y) * z = x * (y * z)$,

ii) there is an element e in G, called the *identity*, such that $e * x = x * e = x$ for all x in G and

iii) for each x in G there is a y in G, such that $x * y = y * x = e$.

We denote y by x^{-1}, the *inverse* of x. We call the group $(G, *)$ *Abelian* if for all x, y in G, $x * y = y * x$. In this case $*$ is usually denoted by $+$, x^{-1} by $-x$, and e by 0. We call $-x$ the *additive inverse* of x. Often the *product* $x * y$ is written simply as xy and x^{-1} is called the *multiplicative inverse* of x.

It turns out that e and x^{-1} are unique. The most familiar examples of Abelian groups are $(G, +)$ with $G = \mathbb{Z}, \mathbb{Q}, \mathbb{R}$ and $\mathbb{C}$. An example of a non-Abelian group is the *general linear group* $GL(n, \mathbb{Z})$ of $n \times n$ matrices with integer entries and determinant ± 1 under matrix multiplication.

A *ring* is set A with at least two distinct elements, denoted by 0 and 1 having two binary operations (addition and multiplication) such that

i) $(A, +)$ is an Abelian group with 0 as its identity,

ii) $1x = x1 = x$ for all x in A and

iii)   the multiplication is associative and *distributive* over the addition:

$$x(y + z) = xy + xz \quad \text{and} \quad (x + y)z = xz + yz.$$

Remark. Some authors don't require that $0 \neq 1$, but we will.

The ring A is *commutative* if $xy = yx$ for all x, y in A. Two important examples of commutative rings are $(\mathbb{Z}, +, \times)$ and the set $\mathbb{Q}[x]$ of polynomials with coefficients in $\mathbb{Q}$ under the usual addition and multiplication of polynomials. An example of a non-commutative ring is the set of 2×2 matrices with the addition and multiplication of matrices. Unless stated otherwise, all rings considered in this book are assumed to be commutative. Another example of such a ring we will meet is the ring $\mathbb{Z}/m\mathbb{Z} = \{0, 1, 2, \dots, m-1\}$ of the *residue classes* of $\mathbb{Z} \bmod m$ ($m > 1$ an integer). To add and multiply the elements of $\mathbb{Z}/m\mathbb{Z}$ we add and multiply them as usual but mod out the sums and products by m, meaning keep only the remainders when divided by m.

A field is a commutative ring K in which every element $\neq 0$ has a multiplicative inverse. The rings $\mathbb{Q}$, $\mathbb{R}$ and $\mathbb{C}$ are fields but not $\mathbb{Z}$. If p is a prime number, $\mathbb{Z}/p\mathbb{Z}$ is a finite field of p elements and is denoted by $\mathbb{F}_p$.

A *module over a* (always commutative) ring A, or simply an A-*module* is an Abelian group $(M, +)$ together with a map $A \times M \ni (a, x) \to ax \in M$, called the *scaling* of x by a, such that for all x, y in M and a, b in A,

i) $a(x + y) = ax + ay$,

ii) $(a + b)x = ax + bx$,

iii) $(ab)x = a(bx)$ and

iv) $1x = x$.

The rings $M = \mathbb{Z}[i]$ and $\mathbb{Z}[\omega]$ of respectively Gaussian and Eisenstein integers are modules over the ring $A = \mathbb{Z}$. In general, if for a prime p $\zeta = \cos\frac{2\pi}{p} + i\sin\frac{2\pi}{p}$ is a primitive p-th root of 1, then $\mathbb{Z}[\zeta] = \{a_0 + a_1\zeta + \cdots + a_{p-1}\zeta^{p-1} \mid a_j \text{ in } \mathbb{Z}\}$ is a $\mathbb{Z}$-module.

To begin with, a field is a ring. A module over a field K is called a *vector space over K*.

A nonempty subset Y of an *object* X (group, ring, field, module, vector space, etc.) is a *subobject* (subgroup, subring, subfield, submodule, subspace, etc.) if Y is closed under the operations for X. For example, if M is a module over a ring A, a non-empty set N of M is a *submodule* of M if for all x, y in N and all a in A, $x + y$ and ax are in N. A subobject Y of X is *proper* if $Y \neq X$. If A is a subring of a ring B, then clearly B is an A-module. Similarly if k is a subfield of a field K, then K is a vector space over k.

Let X, Y be two objects in the same category, i.e. X, Y are either both groups, or both rings, etc. A *homomorphism* (or sometimes simply a *morphism*) is a map $f : X \to Y$ which preserves the operation(s) for the objects in the category. For example, if M and N are both modules over the same ring A, then the map $f : M \to N$ is a *homomorphism of modules* (*linear transformation* or *linear map* if A is a field) if for all x, y in M and all a in A, $f(x+y) = f(x) + f(y)$ and $f(ax) = af(x)$. Two objects X, Y in the same category are *isomorphic* if there is a bijective homomorphism $f : X \to Y$. We write it as $X \cong Y$. A bijective morphism $f : X \to X$ is called an *automorphism* of X.

A ring A may be regarded as a module over itself. In this case, a submodule $\mathfrak{a}$ of A is called an *ideal* of the ring A. (This is precisely how the term ideal was coined by Dedekind in 1870, cf. [11, p. 96].) Every ideal of A contains its zero element.

Often it is possible to divide an object X evenly by its subobjects to get the *quotient* X/Y. We illustrate it with a submodule N of a module M over a ring A. The equivalence relation $\sim$ on M defined by $x \sim y$ if and only if $x - y$ is in N partitions M into disjoint sets, called the *cosets* of M by N, of the form $x + N = \{x + y \mid y \text{ in } N\}$. The following addition and scaling on the set M/N of these cosets turn M/N into a module over A, called the quotient of M by N:

$$(x + N) + (y + N) = (x + y) + N,$$
$$a(x + N) = ax + N.$$

As an example and another way to define the ring $\mathbb{Z}/m\mathbb{Z}$, take $A = \mathbb{Z}$, and $N = m\mathbb{Z}$. Note that N is also an ideal of $\mathbb{Z}$, so we can quotient a ring by its ideals.

Let V be a vector space over a field k. We say that a subset S of V *spans* V if $V = \{c_1 v_1 + \cdots + c_n v_n \mid c_j \text{ in } k, \; v_j \text{ in } S\}$, and V is *finite-dimensional* if there is a finite subset of V which spans V. The *dimension* $\dim_k(V)$ of a finite-dimensional vector space V is the least number of vectors in V that span V. A set $\mathcal{B}$ with a minimal number of vectors that spans V is called a *basis* of V over k.

A field K is called a *field extension* of a field k if k is a subfield of K. If K is a field extension of k, we write it as K/k. If K/k is a field extension, clearly K is a vector space over k. If $\dim_k(K) = n$ is finite, we call K/k a *finite field extension*. We call n the *degree* of the field extension K/k and denote it by $[K : k]$. (The reason for this terminology will become obvious shortly.)

Let M be a $\mathbb{Z}$-module such that there is a finite subset $\{x_1 \ldots, x_r\}$ of M with $M = \{a_1 x_1 + \cdots + a_r x_r \mid a_j \text{ in } \mathbb{Z}\}$. Such a subset with the smallest r is called a $\mathbb{Z}$-*basis* of M. The example that will concern us is the ring $\mathbb{Z}[\zeta]$, ζ a primitive m-th root of unity, considered as a $\mathbb{Z}$-module. One of our goals will be to find an explicit $\mathbb{Z}$-basis for $\mathbb{Z}[\zeta]$.

2.2   Galois Extensions

We assume all our fields to be subfields of $\mathbb{C}$. Let K/k be a field extension. The set $\mathrm{Gal}(K/k)$ of the field automorphisms σ of K such that $\sigma(a) = a$ for all a in k is a (usually non-Abelian) group under the composition of maps. It is called the *Galois group* of K over k. In general, for a finite extension K/k, $|\mathrm{Gal}(K/k)| \leq [K:k]$. We call K/k a *Galois extension* if the equality holds.

Examples of Galois Groups

First, let K/k be any field extension, not necessarily finite. Let α in K be a root of a polynomial

$$f(x) = c_0 + c_1 x + \cdots + c_n x^n$$

over k. If $\sigma \in \mathrm{Gal}(K/k)$, then

$$f(\sigma(\alpha)) = c_0 + c_1 \sigma(\alpha) + \cdots + c_n (\sigma(\alpha))^n$$
$$= \sigma(f(\alpha)) = \sigma(0) = 0.$$

Thus $\sigma(\alpha)$ is also a root of $f(x)$. This simple observation will be crucial to what follows.

Let K be a quadratic field, a field extension of $\mathbb{Q}$ of degree 2. Then one checks that (Exercise 16) $K = \mathbb{Q}(\sqrt{d}) = \{r + s\sqrt{d} \mid r, s \in \mathbb{Q}\}$ for a square-free integer $d \neq 0, 1$.

Example 2.1. Let us take $d = -1$. There are exactly two automorphisms of K whose restrictions to $\mathbb{Q}$ is the identity map on $\mathbb{Q}$. The identity map 1 on K itself and σ which takes i to its conjugate, the other root $-i$ of $x^2 + 1$. Thus $\mathrm{Gal}(K/k) \cong \{\pm 1\}$ and $\mathbb{Q}(i)$ is a Galois extension of $\mathbb{Q}$.

Example 2.2. Now take $d = -3$. Then $\mathbb{Q}(\omega) = \{r + s\omega \mid r, s \in \mathbb{Q}\}$. The Galois group $\mathrm{Gal}(K/k)$ consists of two elements, the identity automorphism 1 of K and the automorphism σ of K such that $\sigma(\omega) = \bar{\omega}$. [Note that $\bar{\omega} = \omega^2 = \frac{1}{\omega}$.] Hence $\mathbb{Q}(\omega)/\mathbb{Q}$ is also an Abelian extension.

Example 2.3. Let α be the real cube root of 2, $\alpha = \sqrt[3]{2}$, $K = \mathbb{Q}(\alpha)$ the smallest subfield of $\mathbb{C}$ containing α. The other cube roots of 2 which are $\omega\alpha$ and $\omega^2\alpha$ are not in K. Thus there is only one element in the Galois group Gal($K/\mathbb{Q}$), namely the identity element of the group Gal($K/\mathbb{Q}$). Since $[K : \mathbb{Q}] = 3$ but $|\,\text{Gal}(K/\mathbb{Q})| = 1$, the extension $K/\mathbb{Q}$ is not Galois.

The following is a standard result in field theory:

Theorem 2.4. *If K/k is a field extension of degree d, then there is an α in K such that $1, \alpha, \alpha^2, \ldots, \alpha^{d-1}$ is a basis of K as a vector space over k.*

In fact, α is a root of an irreducible polynomial $f(x)$ over k of degree d.

Definition 2.5. *If all the d roots of this $f(x)$ are in K, we call the extension K/k normal.*

Remark 2.6. i) According to our definition of the Galois extension, an extension is normal if and only if it is Galois.

ii) The Galois group Gal(K/k) is often defined only for normal extensions, in which case Gal(K/k) is always equal to the degree $[K : k]$ of the field extension K/k.

2.3   Integral Domains

A nonzero element a of a ring A (always commutative) is called a *zero divisor* if $ab = 0$ for a nonzero b in A. In the ring $\mathbb{Z}/6\mathbb{Z}$, 2, 3, and 4 are the only divisors of zero. A field has no divisor of zero. A ring without zero divisors is called an *integral domain* or simply a *domain*. We have already discussed many integral domains which are not fields, e.g. $\mathbb{Z}$, $\mathbb{Z}[i]$, $\mathbb{Z}[\omega]$ and $\mathbb{Z}[\sqrt{d}]$ for $d \neq 0$, a square-free integer, which are relevant to our subject.

An element u in A is a *unit* if $uv = 1$ for some v in B. For example, the only units in the ring $\mathbb{Z}$ are ± 1.

Definition 2.7. A domain A is a *Euclidean domain* if there is a map which assigns to each nonzero element α of A a non-negative integer $d(\alpha)$ such that for all nonzero α, β in A,

i) $d(\alpha) \leq d(\alpha\beta)$, and

ii) A has elements q (the *quotient*) and γ (the *remainder*) so that $\alpha = q\beta + \gamma$ and either $\gamma = 0$ or $d(\gamma) < d(\beta)$.

With the Euclidean algorithm, both $\mathbb{Z}$ and the ring $k[x]$ of polynomials over a field k are Euclidean domains. For $\mathbb{Z}$, $d(\alpha) = |\alpha|$ and for $k[x]$, $d(f(x)) = \deg f(x)$.

Example 2.8. For an $\alpha = a + bi$ in the field $\mathbb{Q}[i]$, the *conjugate* of α is the element $\bar{\alpha} = a - bi$ of $\mathbb{Q}[i]$. The *norm* of α is the rational number $N(\alpha) = \alpha\bar{\alpha} = a^2 + b^2$ which is non-negative and $= 0$ if and only if $\alpha = 0$. Moreover, $N(\alpha\beta) = N(\alpha)N(\beta)$. We show that the ring $\mathbb{Z}[i]$ is *norm Euclidean*, i.e. $d(\alpha) = N(\alpha)$ makes $\mathbb{Z}[i]$ a Euclidean domain.

The condition i) in the definition is obvious. For ii) let $\alpha = a + ib$, $\beta = c + id$ be in $\mathbb{Z}[i]$. Then

$$\frac{\alpha}{\beta} = \frac{ac + bd}{c^2 + d^2} + \frac{bc - ad}{c^2 + d^2} i$$
$$= A + iB, \text{ say.}$$

Note that A and B are in $\mathbb{Q}$, and not necessarily in $\mathbb{Z}$.

Choose integers x and y so that $|A - x| \leq \frac{1}{2}$ and $|B - y| \leq \frac{1}{2}$. If we put $q = x + iy$, a quick calculation shows that $N\left(\frac{\alpha}{\beta} - q\right) < 1$. We take γ to be $\alpha - q\beta$. If $\gamma \neq 0$, then $N(\gamma) = N\left(\beta\left(\frac{\alpha}{\beta} - q\right)\right) < N(\beta)$. ∎

For a domain A and nonzero a, b in A, b is a *divisor* of a if $a = bc$ for some c in A. We also say that b is a *factor* of a or a is a *multiple* of b. A nonzero and non-unit element π of A is a *prime element* of A if the only divisors of π are $u\pi$ with u a unit in A.

Definition 2.9. A domain A is a *unique factorization domain* (UFD for short) if each nonzero element of A which is not a unit is, up to a unit factor, a unique product of powers of distinct primes. A domain is a *principal ideal domain* (or PID) if each of its ideal $\mathfrak{a}$ is *principal*, i.e. is of the form $\mathfrak{a} = aA = \{ax \mid x \in A\}$.

Theorem 2.10. *Every Euclidean domain is a PID.*

Theorem 2.11. *Every PID is a UFD.*

Thus every Euclidean domain is a UFD. In particular, the ring $k[x]$ is a UFD. The following is a well-known result, but it will not concern us.

Theorem 2.12. *For $d > 1$, the ring $\mathbb{Z}[\sqrt{d}]$ is norm Euclidean $\Leftrightarrow d = 2, 3, 5, 6, 7, 11, 13, 17, 19, 21, 29, 33, 37, 41, 55$ and 73.*

Let M be a module over a commutative ring A with 1. We say that M is a *finitely generated* A-module if there is a finite set $\{x_1, \ldots, x_r\}$ of elements of M such that each x in M is a linear combination

$$x = c_1 x_1 + \cdots + c_r x_r \tag{2.1}$$

for some c_j in A. A finitely generated module M over A is *free* of rank r if

$$M = Ax_1 \oplus \cdots \oplus Ax_r. \tag{2.2}$$

This means that the representation (2.1) is unique for each x in M.

The following fact from an undergraduate course in abstract algebra will be recalled when needed. For a proof, see [23].

Theorem 2.13. *A finitely generated module over a PID is free.*

2.4 Factoring Rational Primes in $\mathbb{Z}[i]$

Let A be the ring $\mathbb{Z}[i]$ of Gaussian integers and $p = 2, 3, 4, \ldots$ a rational prime. This p may or may not be a prime element of A. To find exactly when it is, recall the famous theorem of Fermat on the sum of two squares, which was proved by Euler (cf. [8, p. 48]).

Theorem 2.14 (Fermat). *An odd prime p in $\mathbb{Z}$ is a sum of two squares $(p = a^2 + b^2)$ if and only if $p = 4k + 1$ for k in $\mathbb{N}$.*

The norm of any divisor of $\alpha = a + ib$ must be a divisor of $N(\alpha) = a^2 + b^2$, and for $\alpha = \beta\gamma$ with β, γ both non-units, $1 < N(\beta) < N(\alpha)$ (only the units have norm 1). Therefore, if $a^2 + b^2$ is a prime, then α has to be a prime in $\mathbb{Z}[i]$. We have thus proved the following fact:

Theorem 2.15. *A prime p is a sum of two squares, $p = a^2 + b^2 \Leftrightarrow p$ is a product $(a + ib)(a - ib)$ of two primes $a \pm ib$ in $\mathbb{Z}[i]$.*

For $p = 2$, its two prime factors $1+i$, $1-i$ in $\mathbb{Z}[i]$ are associates: $1+i = i(1-i)$. Therefore,
$$2 = i(1 - i)^2.$$
We say that 2 *ramifies* in $\mathbb{Z}[i]$. By Fermat's Theorem (Theorem 2.15), $p \equiv 1 \pmod 4 \Leftrightarrow p$ is a product
$$p = \pi_1 \pi_2$$
of two primes π_1, π_2 in $\mathbb{Z}[i]$. Moreover, π_1 and π_2 are complex conjugates of each other and hence they are distinct. This discussion can be wrapped up as follows: In order to do that, observe that $\{1, i\}$ is a $\mathbb{Z}$-bases of $\mathbb{Z}[i]$ and so is its conjugate $\{1, -i\}$. These two bases make a 2×2 matrix
$$A = \begin{pmatrix} 1 & i \\ 1 & -i \end{pmatrix}$$
with $|\det(A)| = 2$, called the *discriminant* of $\mathbb{Q}(i)$.

Theorem 2.16. *Let p be a prime. Then*

i) p ramifies in $\mathbb{Z}[i] \Leftrightarrow$ it divides the discriminant of $\mathbb{Q}(i)$,

ii) p factors into two distinct primes of $\mathbb{Z}[i] \Leftrightarrow p \equiv 1 \pmod 4$, *and*

iii) p stays prime in $\mathbb{Z}[i] \Leftrightarrow p \equiv 3 \pmod 4$.

EXERCISES

1. As a prelude to Dirichlet's Theorem on Units, let $d > 1$ be a square-free integer. Let G_d be the set of integer solutions (x, y) of the *Pell equation*

$$x^2 - dy^2 = 1. \tag{2.3}$$

Define a binary operation $*$ on G_d by

$$(x_1, y_1) * (x_2, y_2) = (x_1 x_2 + d y_1 y_2, x_1 y_2 + x_2 y_1).$$

Show that $(G_d, *)$ is an Abelian group with identity $(1, 0)$ and $(x, y)^{-1} = (x, -y)$.

2. With d as in 1, let G be the group of 2×2 matrices $\begin{pmatrix} x & dy \\ y & x \end{pmatrix}$ of determinant 1 with x, y integers, and $G' = (\mathbb{Z}[\sqrt{d}])^{\times}$, the group of units in the ring $\mathbb{Z}[\sqrt{d}]$ of positive norm.

   Show that G, G', and G_d are all isomorphic to each other. In other words, the same group has been obtained in three different ways.

3. It will follow from the Dirichlet Theorem that $G_d \cong \{\pm 1\} \times \mathbb{Z}$. The elements of G_d with $x > 0$ are points on the parabola $x^2 - dy^2 = 1$ (cf. Figure 2.1) in the right half of the plane. Those with $x < 0$ come from -1 in the factor group $\{\pm 1\}$ of G_d.

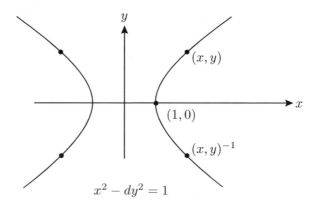

FIGURE 2.1: Pell equation as hyperbola.

There are four generators of the free part of G_d. To find one, put $y = 1, 2, 3, \ldots$ until it becomes a square x_1^2. (Dirichlet's Theorem guarantees that sooner or later it will.) Find (x_1, y_1) with $x_1, y_1 > 0$ for G_d if $d = 2, 3, 15, 67, 77, 94$.

4. i) Use the group law in G_d to compute $(x_5, y_5) = (x_1, y_1)^5$ for $d = 2$.

    ii) Rewriting the Pell equation as

    $$\left(\frac{x}{y}\right)^2 = d + \frac{1}{y^2}$$

    shows that $\frac{x}{y}$ approximates $\sqrt{d}$ if y is very large. Use part i) to find a rational approximation to $\sqrt{2}$.

    iii) Compare this approximation with the one given by a computer.

5. Show that $\mathbb{Z}[\sqrt{d}]$ and $\mathbb{Z}[\omega]$ are rings.

6. In Definition 2.7 of a Euclidean domain, show that the equality $d(\alpha) = d(\alpha\beta)$ holds if and only if β is a unit.

7. Show that the only units of $\mathbb{Z}[\omega]$ are $\pm 1, \pm\omega, \pm\omega^2$.

8. Show that the ring $\mathbb{Z}[\sqrt{-2}]$ is norm Euclidean.

9. Show that $\mathbb{Z}[\omega]$ is norm Euclidean.

10. Show that a Euclidean domain A is a PID.

    Hint: If $\mathfrak{a}$ is a nonzero ideal of A, choose an element $\alpha \neq 0$ in A with the smallest $d(\alpha)$. Show that $\mathfrak{a} = (\alpha)$.

11. Give an example of a UFD which is not a PID.

12. Let $\mathfrak{m}$ be an ideal of a ring A. We call $\mathfrak{m}$ *maximal* if $\mathfrak{m} \neq A$ and there is no ideal $\mathfrak{a}$ with $\mathfrak{m} \subsetneq \mathfrak{a} \subsetneq A$. We call an ideal $\mathfrak{p}$ of a ring A *prime* if $ab \in \mathfrak{p}$ implies either $a \in \mathfrak{p}$ or $b \in \mathfrak{p}$.

    Show that a maximal ideal is a prime ideal but not conversely.

13. Show that $\mathfrak{m}$ is a maximal ideal of $A \Leftrightarrow A/\mathfrak{m}$ is a field and $\mathfrak{p}$ a prime ideal of $A \Leftrightarrow A/\mathfrak{p}$ is a domain.

14. If $f : A \to B$ is a ring homomorphism, show that

    i) its *kernel* $\mathrm{Ker}(f) = \{x \in A \mid f(x) = 0\}$ is an ideal of A,

    ii) f is injective $\Leftrightarrow \mathrm{Ker}(f) = \{0\}$ and

    iii) if A is a field, then f is either the zero map or injective.

15. *First Isomorphism Theorem*

Let X, Y be both Abelian groups, both commutative rings with 1, or both modules over a commutative ring with 1 (vector spaces if this ring is a field). Let $f : X \to Y$ be a morphism of these two objects (in the same category). Let $\mathrm{Ker}(f) = \{x \in X \mid f(x) = 0\}$ be the *kernel* of f whereas $f(X) = \{f(x) \mid x \in X\}$ its *image*. Show that $X/\mathrm{Ker}(f) \cong f(X)$.

16. Show that any quadratic extension of $\mathbb{Q}$ is $\mathbb{Q}(\sqrt{d})$ for a square free integer $d \neq 0, 1$.

17. If ℓ is an odd prime and $\zeta = \zeta_\ell = \cos \frac{2\pi}{\ell} + i \sin \frac{2\pi}{\ell}$, show that $\mathbb{Q}(\zeta) = \{a_0 + a_1\zeta + a_2\zeta^2 + \cdots + a_{\ell-1}\zeta^{\ell-1} \mid a_j \in \mathbb{Q}\}$ is a field extension of $\mathbb{Q}$ of degree $p - 1$.

18. Show that $\mathrm{Gal}(\mathbb{Q}(\zeta)/\mathbb{Q})$ is isomorphic to the group of units of the ring $\mathbb{Z}/p\mathbb{Z}$. Thus $\mathbb{Q}(\zeta)/\mathbb{Q}$ is a Galois extension.

    Hint: If $\sigma \in \mathrm{Gal}(\mathbb{Q}(\zeta)/\mathbb{Q})$, then $\sigma(\zeta)$ is another root of unity.

19. Let m be any positive integer, not necessarily a prime, and $\zeta = \cos \frac{2\pi}{m} + i \sin \frac{2\pi}{m}$.

    Show that $\mathbb{Q}(\zeta)/\mathbb{Q}$ is still Galois. What is its Galois group?

3

Basic Concepts

3.1 Generalities

Let K/k be a field extension and suppose α is an element of K. We say that α is *algebraic over* k if α satisfies a nonzero polynomial over k. Suppose $n = \dim_k(K)$ is finite and α is in K. Then the $n+1$ vectors $1, \alpha, \ldots, \alpha^n$ cannot be linearly independent and hence satisfy a nontrivial linear relation

$$c_0 + c_1\alpha + \cdots + c_n\alpha^n = 0,$$

with c_j in k. This not only shows that α is algebraic over k but also proves that it is a root of a nonzero polynomial of degree at most n over k. The smallest degree of a polynomial over k satisfied by α is called the *degree of α over k*. It is denoted by $\deg_k(\alpha)$.

3.2 Algebraic Integers

The subject of algebraic number theory originated with Gauss, who studied the arithmetic in the ring $\mathbb{Z}[i] = \{x + iy | x, y \in \mathbb{Z}\}$ of the so called *Gaussian integers*. We begin with a useful fact about field extensions which is true for the ones to be dealt with in this book.

Definition 3.1. A field extension K/k is a *simple extension* if there is an element α in K such that $K = k(\alpha)$.

Here $k(\alpha)$ is the field of all quotients of polynomials in α over k. It is the smallest field containing k and α. We say that K has been *obtained by adjoining α to k*. We also say that α *generates K over k*.

From now on, we shall regard $\mathbb{C}$, the field of complex numbers, as our *universal domain*. This essentially means that all fields, unless stated otherwise, shall be subfields of $\mathbb{C}$. A field must have at least two distinct elements, namely, 0 and 1. Therefore, a subfield of $\mathbb{C}$ must contain $\mathbb{Z}$, and hence it must be an extension of $\mathbb{Q}$. The following is a standard result from field theory (cf. [8, p. 72]).

Theorem 3.2. *If k is a subfield of $\mathbb{C}$, then any finite extension K/k is a simple extension.*

Definition 3.3. A *number field* is a finite extension of $\mathbb{Q}$. A number field K is a *quadratic field* or a *cubic field* according as $[K : \mathbb{Q}]$ is 2 or 3. We call a field extension K/k an *extension of number fields* if k is a subfield of the number field K. Clearly, k is also a number field.

Definition 3.4. A complex number α is an *algebraic number* if it is algebraic over $\mathbb{Q}$.

It is not hard to see that the set of all algebraic numbers is a subfield of $\mathbb{C}$. It is called the *algebraic closure of* $\mathbb{Q}$ *in* $\mathbb{C}$ and is denoted by $\overline{\mathbb{Q}}$.

Every element of a number field K with $[K : \mathbb{Q}] = n$ is an algebraic number of degree at most n. By Theorem 3.2, there is always an α in K with $\deg_{\mathbb{Q}}(\alpha) = n$.

The following definition is crucial to what follows.

Definition 3.5. A polynomial is *monic* if its leading coefficient is 1.

Note that since an algebraic number satisfies a polynomial $f(x)$ over $\mathbb{Q}$, by clearing the denominators from the coefficients of $f(x)$ if necessary, we may assume that the coefficients of $f(x)$ are actually in $\mathbb{Z}$. But then $f(x)$ may not be monic. In other words, an algebraic number may not be a root of a monic polynomial over $\mathbb{Z}$. The algebraic numbers which are roots of monic polynomials over $\mathbb{Z}$ play a central role in algebraic number theory and have a special name.

Definition 3.6. An algebraic number is an *algebraic integer* if it is a root of a monic polynomial over $\mathbb{Z}$.

Theorem 3.7. *The set of algebraic integers is a subring of $\mathbb{C}$ containing $\mathbb{Z}$ as a subring.*

We postpone the proof of this theorem, and denote the ring of all algebraic integers by $\mathcal{O}$. It follows that for a field K the set $\mathcal{O} \cap K$ is a subring of K. It is denoted by $\mathcal{O}_K$. For a number field K, the ring $\mathcal{O}_K$ has the same relation to K as $\mathbb{Z}$ has to $\mathbb{Q}$.

Definition 3.8. The ring $\mathcal{O}_K$ is called the *ring of integers of K*.

EXERCISE

Prove that the ring of integers of $\mathbb{Q}$ is $\mathbb{Z}$.

[*Hint*: To prove this plug $\alpha = a/b$ in $\mathbb{Q}$ (taken in the lowest form) into the polynomial equation defining an algebraic integer. Show that b must divide 1.]

Theorem 3.7 follows at once from a standard proposition in commutative algebra, which we shall prove after some preliminary remarks. Recall that for a commutative ring A with 1, the definition of a *module over A*, or simply an *A-module*, is the same as that of a vector space except that the scalars are taken from the ring A, which may not be a field. A module M over A is *finitely generated* or simply a *finite A-module* if M contains a finite set S which generates M. This means that M consists of linear combinations of elements of S with coefficients in A. A commutative ring A with 1 is an *integral domain*, or for short a *domain*, if it has no nonzero *divisors of zero*, i.e. if $x, y \in A$ and $xy = 0$ implies that either $x = 0$ or $y = 0$. All subrings are assumed to contain 1.

Definition 3.9. Let A, B be two domains. Suppose A is a subring of B. An element of B is *integral over A* if it is a root of a monic polynomial over A.

Proposition 3.10. *Let a domain A be a subring of a domain B and $\alpha \in B$. The following are equivalent:*

(1) α is integral over A.

(2) The ring $A[\alpha]$ of polynomials in α over A is a finite A-module.

(3) There is a finite A-module $M \neq \{0\}$ such that $\alpha M \subseteq M$.

Proof. (1) implies (2): If

$$\alpha^n + c_{n-1}\alpha^{n-1} + \cdots + c_0 = 0 \tag{3.1}$$

with c_j in A, we put

$$M = A + A\alpha + \cdots + A\alpha^{n-1}.$$

Since by equation (3.1), n-th and higher powers of α can be expressed as linear combinations of the lower powers of α with coefficients in A, it is clear that $A[\alpha] = M$, and the A-module M is generated by the finitely many elements $1, \alpha, \ldots, \alpha^{n-1}$.

(2) implies (3): The finitely generated A-module $M = A[\alpha] \neq \{0\}$ and $\alpha M \subseteq M$.

(3) implies (1): Let $M = A\alpha_1 + \cdots + A\alpha_n \neq \{0\}$. Then $\alpha\alpha_i = \sum_{j=1}^{n} a_{ij}\alpha_j$, or

$$\alpha \begin{pmatrix} \alpha_1 \\ \vdots \\ \alpha_n \end{pmatrix} = T \begin{pmatrix} \alpha_1 \\ \vdots \\ \alpha_n \end{pmatrix}, \quad \text{i.e.} \quad (\alpha I - T) \begin{pmatrix} \alpha_1 \\ \vdots \\ \alpha_n \end{pmatrix} = 0,$$

where T is the n by n matrix (a_{ij}) with entries in A and I the n by n identity matrix. Since the vector

$$\begin{pmatrix} \alpha_1 \\ \vdots \\ \alpha_n \end{pmatrix} \neq 0,$$

the determinant $\det(\alpha I - T) = 0$, which shows that α satisfies a monic polynomial over A. $\qquad \square$

Proof of Theorem 3.7. Let α and β be algebraic integers. Then $\mathbb{Z}[\alpha]$ and $\mathbb{Z}[\beta]$ are finitely generated $\mathbb{Z}$-modules, say generated by the finite sets $\{\alpha_i\}$ and $\{\beta_j\}$, respectively. Moreover, $M = \mathbb{Z}[\alpha, \beta]$ is not zero and is generated as a $\mathbb{Z}$-module by $\{\alpha_i\beta_j\}$. Now if $\gamma = \alpha \pm \beta$ or $\alpha\beta$, it is clear that $\gamma M \subseteq M$. By Proposition 3.10, γ is an algebraic integer. This shows that $\mathcal{O}$ is closed under ring operations and hence it is a subring of $\mathbb{C}$. Since 0 and 1 are algebraic integers, $\mathbb{Z} \subseteq \mathcal{O}$. $\qquad \square$

Theorem 3.11. *If β is an algebraic number, then there is an algebraic integer α and a rational integer m (that is an m in $\mathbb{Z}$) such that $\beta = \alpha/m$.*

Proof. Suppose

$$c_n\beta^n + \cdots + c_1\beta + c_0 = 0 \tag{3.2}$$

with c_j in $\mathbb{Z}$ and $c_n \neq 0$. Let $m = c_n$. Multiplying the equation (3.2) throughout by m^{n-1}, we obtain

$$(m\beta)^n + \cdots + m^{n-2}c_1(m\beta) + m^{n-1}c_0 = 0.$$

This shows that $\alpha = m\beta$ is a root of a monic polynomial over $\mathbb{Z}$ and hence is an algebraic integer α, which proves the theorem. $\qquad \square$

Corollary 3.12. *Any given number field K is the field of fractions of its ring of integers $\mathcal{O}_K$.*

3.3   Integral Bases

Let A be a commutative ring with 1. Suppose $M \neq \{0\}$ is an A-module. We say that M is a *free A-module of rank n* (n being an integer ≥ 1) if there are n elements $\alpha_1, \ldots, \alpha_n$ in M such that every element α of M can be uniquely written as

$$\alpha = a_1\alpha_1 + \cdots + a_n\alpha_n$$

with a_j in A. We write it as

$$M = A\alpha_1 \oplus \ldots \oplus A\alpha_n.$$

The set $\{\alpha_1, \ldots, \alpha_n\}$ is called a *basis* of M *over A*. If the elements of a basis are taken in a fixed order, it is called an *ordered basis*. In this section, we shall prove that for a number field K, its ring of integers $\mathcal{O}_K$ is a free $\mathbb{Z}$-module of rank $[K : k]$. We recall some basic facts needed from linear algebra and Galois theory.

Suppose $x = (x_{ij})$ is in $M(n, A)$, that is x is an n by n matrix with entries in A.

Definition 3.13. The *trace* $\mathrm{tr}(x)$ of x is the sum $x_{11} + \cdots + x_{nn}$ of the diagonal entries of x.

The following theorem is obvious.

Theorem 3.14. *Let x, y be in $M(n,A)$ and a in A. Then*

(1) $\mathrm{tr}(x + y) = \mathrm{tr}(x) + \mathrm{tr}(y)$.

(2) $\mathrm{tr}(ax) = a\,\mathrm{tr}(x)$

(3) $\mathrm{tr}(xy) = \mathrm{tr}(yx)$.

Now suppose M a free A-module of rank n over A. Let $\lambda : M \to M$ be a homomorphism of A-modules, or simply an *A-homomorphism*. We associate a matrix L over A to λ with respect to an ordered basis $\{\alpha_1, \ldots, \alpha_n\}$ of M over A in the same way as to a linear transformation. If L_1 and L_2 are the matrices of λ with respect to two ordered bases, then $L_2 = P^{-1}L_1P$ for some P in $GL(n, A)$, that is, for a matrix P over A whose determinant has multiplicative inverse in A.

For the rest of the section, let K/k be an extension of number fields. Since the dimension $\dim_k(K)$ cannot be more than $\dim_{\mathbb{Q}}(K)$, it is clear that K/k is a finite extension. We may regard K as a k-module of rank $n = [K : k]$. For α in K, the multiplication by α is a k-homomorphism $m_\alpha : K \to K$. Let L_α be the matrix of m_α with respect to an ordered basis of K over k.

Definition 3.15. The *relative norm* and the *relative trace* for K/k of α are

$$N_{K/k}(\alpha) = \det(L_\alpha)$$

$$\mathrm{tr}_{K/k}(\alpha) = \mathrm{tr}(L_\alpha).$$

We know from linear algebra that $\det(P^{-1}LP) = \det(L)$ and $\mathrm{tr}(P^{-1}LP) = \mathrm{tr}(L)$. Therefore, these terms do not depend on the ordered basis chosen to define L_α and hence, are well defined. If there is no likelihood of confusion, we omit the adjective "relative",

Remark 3.16. The norm and the trace are maps from K to k.

The following is obvious.

Theorem 3.17. *For α, β in K and a, b in k,*

   (1) $N_{K/k}(\alpha\beta) = N_{K/k}(\alpha)N_{K/k}(\beta)$

   (2) $\mathrm{tr}_{K/k}(a\alpha + b\beta) = a\ \mathrm{tr}_{K/k}(\alpha) + b\ \mathrm{tr}_{K/k}(\beta).$

In particular,

   (1) $N_{K/k}(a) = a^n$.

   (2) $\mathrm{tr}_{K/k}(a) = na.$

By Theorem 3.2, any extension K/k of number fields is a simple extension. Let $K = k(\alpha)$. If $f(x)$ is the *minimal polynomial of α over k*, that is, the monic polynomial of the smallest degree over k of which α is a root, then clearly $f(x)$ is irreducible over k and $\deg_k(\alpha) \le [K : k]$. Let $\deg f(x) = n$.

EXERCISES

1. Show that $f(x)$ has n distinct roots.

    [*Hint*: If α is a repeated root of $f(x)$, then $f(x) = (x - \alpha)^2 g(x)$ for some $g(x)$. Differentiate this equation and put $x = \alpha$ in the resulting equation to get a contradiction.]

2. Show that the field $k(\alpha) = $ the ring $k[\alpha]$ of polynomials in α over k.

    [*Hint*: If $g(\alpha)$ is not zero, then $f(x)$ and $g(x)$ are coprime (why?). Show that (by the Euclidean algorithm) one can explicitly write $1 = a(x)g(x) + b(x)f(x)$ for some $a(x)$ and $g(x)$ (how?). Put in this equation $x = \alpha$ to show that $g(\alpha)$ has a multiplicative inverse in $k[\alpha]$.]

Thus $[K : k] \leq \deg f(x) \leq [K : k]$, which shows that $[K : k] = \deg f(x) = n$. Hence, we obtain the following result.

Theorem 3.18. *For any extension K/k of number fields, there is an α in K, such that $K = k[\alpha]$ and*

$$[K : k] = \deg_k(\alpha).$$

Remark 3.19. This explains why the dimension $\dim_k(K)$, which we denoted by $[K : k]$ is called the degree of K over k. It is the degree $\deg_k(\alpha)$ of the minimal polynomial of α that generates K over k.

Remark 3.20. Suppose $\sigma : K \to \mathbb{C}$ is a ring homomorphism. Since by definition, $\sigma(1) = 1$, the restriction $\sigma_{|\mathbb{Q}} =$ the identity map on $\mathbb{Q}$. Thus a ring homomorphism from K to $\mathbb{C}$ is a $\mathbb{Q}$-homomorphism of $\mathbb{Q}$-modules from K to $\mathbb{C}$. In general, suppose K/k is an extension of number fields. We call a ring homomorphism $\sigma : K \to \mathbb{C}$ a k-*homomorphism* if the restriction $\sigma_{|k}$ is the identity map.

Theorem 3.21. *There are exactly $[K : k]$ k-homomorphisms $\sigma : K \to \mathbb{C}$.*

Proof. Such a σ fixes k, hence $\sigma(f(\alpha)) = f(\sigma(\alpha)) = 0$, which shows that $\sigma(\alpha)$ is also a root of $f(x)$. Therefore, in view of $K = k(\alpha)$, σ is determined by the value $\sigma(\alpha)$, which can be any of the n distinct roots of $f(x)$. This proves the theorem.

Let $\sigma_1, \ldots, \sigma_n$ be these $n = [K : k]$ k-homomorphisms. By Galois Theory, $k = \{\alpha \in K \mid \sigma_j(\alpha) = \alpha, \ j = 1, \ldots, n\}$. $\qquad\square$

Theorem 3.22. *If α is an algebraic integer, then for every i, $\sigma_i(\alpha)$ is also an algebraic integer.*

Proof. Let $f(x)$ be a monic polynomial over $\mathbb{Z}$ satisfied by α. Since σ_i fixes $\mathbb{Q}$, we have $f(\sigma_i(\alpha)) = \sigma_i(f(\alpha)) = 0$. So $\sigma_i(\alpha)$ also satisfies the same monic polynomial over $\mathbb{Z}$. $\qquad\square$

To prove that $\mathcal{O}_K$ has a $\mathbb{Z}$-basis, we begin with

Theorem 3.23. *The set $\{\alpha_1, \ldots, \alpha_n\} \subseteq K$ is a basis of K over k if and only if the matrix $P = (\sigma_i(\alpha_j))$ is non-singular.*

Proof. Let us prove the theorem first for a set of the form $\{1, \alpha, \ldots, \alpha^{n-1}\}$ with α in K. Since K/k is a simple extension, such a basis exists. In this case $\det(P)$ is the well-known van der Monde determinant, which can be simplified to

$$\prod_{i<j}(\sigma_i(\alpha) - \sigma_j(\alpha)).$$

Therefore, $\det(P)$ is nonzero if and only if $\sigma_1(\alpha), \ldots, \sigma_n(\alpha)$ are all distinct. This happens if and only if the minimal polynomial of α over k has n (distinct) roots $\Leftrightarrow \deg_k(\alpha) = n \Leftrightarrow \{1, \alpha, \ldots, \alpha^{n-1}\}$ is a basis of K over k.

For any two bases of K over k, the corresponding P are conjugate by a non-singular matrix. Therefore, if the theorem is true for one set $\{\alpha_1, \ldots, \alpha_n\}$, it is true for all of them. $\square$

Theorem 3.24. *For α in K,*

$$\operatorname{tr}_{K/k}(\alpha) = \sum_{j=1}^{n} \sigma_j(\alpha),$$

$$N_{K/k}(\alpha) = \prod_{j=1}^{n} \sigma_j(\alpha).$$

Proof. Let $\{\alpha_1, \ldots, \alpha_n\}$ be a basis of K over k and write

$$\alpha \alpha_j = \sum_{r=1}^{n} a_{rj} \alpha_r$$

with a_{rj} in $\mathbb{Q}$. Apply the k-homomorphism σ_i to this equation to get

$$\sigma_i(\alpha)\sigma_i(\alpha_j) = \sum_{r=1}^{n} \sigma_i(\alpha_r) a_{rj},$$

which can be written as a single matrix equation $DP = PL_\alpha$, or

$$D = PL_\alpha P^{-1},$$

where $P = (\sigma_i(\alpha_j))$ and D is the diagonal matrix $\operatorname{diag}(\sigma_1(\alpha), \ldots, \sigma_n(\alpha))$. The theorem follows by taking the norm and the trace of the matrix D above. $\square$

Now the following is almost obvious.

Theorem 3.25. *For α in $\mathcal{O}_K$, the norm and the trace of α over $K/\mathbb{Q}$ are in $\mathbb{Z}$.*

Proof. By Remark 3.16, the norm and the trace are in $\mathbb{Q}$, whereas by Theorem 3.22 and Theorem 3.24, they are in $\mathcal{O}_K$. Hence they are in $\mathbb{Q} \cap \mathcal{O}_K = \mathbb{Z}$. (See Exercise 3.2). $\square$

Now we prove the important theorem we mentioned earlier.

Theorem 3.26. *Suppose K is a number field with $[K : \mathbb{Q}] = n$. There are algebraic integers $\alpha_1, \ldots, \alpha_n$, such that*

$$\mathcal{O}_K = \mathbb{Z}\alpha_1 \oplus \cdots \oplus \mathbb{Z}\alpha_n.$$

Proof. By Theorem 3.11, we can choose a basis $\{\alpha_1, \ldots, \alpha_n\}$ of K over $\mathbb{Q}$ consisting of algebraic integers. Let P be the n by n matrix $(\sigma_i(\alpha_j))$. Then, being fixed by every σ_i,

$$\Delta(\alpha_1, \ldots, \alpha_n) = \det(P^2) \qquad (3.3)$$

is a nonzero integer. Choose such a basis with minimum $|\Delta(\alpha_1, \ldots, \alpha_n)|$. Now suppose $\alpha \in \mathcal{O}_K$ and

$$\alpha = a_1\alpha_1 + \cdots + a_n\alpha_n$$

with $a_j \in \mathbb{Q}$. If an a_j, say a_1 is not in $\mathbb{Z}$, we can write

$$a_1 = a + r \quad (0 < r < 1).$$

Put

$$\omega_1 = \alpha - a\alpha_1 = r\alpha_1 + a_2\alpha_2 + \cdots + a_n\alpha_n$$

and

$$\omega_j = \alpha_j$$

for $j \geq 2$.

It is clear that $\{\omega_1, \ldots, \omega_n\}$ is also a basis consisting of algebraic integers. The transition matrix for these two bases is

$$\left(\begin{array}{c|c} r & 0 \\ \hline a_2 & \\ \vdots & I_{n-1} \\ a_n & \end{array} \right)$$

which gives $\Delta(\omega_1, \ldots, \omega_n) = r^2\Delta(\alpha_1, \ldots, \alpha_n)$, contradicting the minimality of $|\Delta(\alpha_1, \ldots, \alpha_n)|$. $\qquad \square$

Definition 3.27. The set $\{\alpha_1, \ldots, \alpha_n\}$ in Theorem 3.26 is called an *integral basis* of the ring $\mathcal{O}_K$ (of integers of K), or for brevity, an *integral basis* of K.

Remark 3.28. Any two integral bases are connected by an n by n matrix P in $GL(n, \mathbb{Z})$, i.e. an invertible matrix P over $\mathbb{Z}$, such that P^{-1} also has entries in $\mathbb{Z}$.

Definition 3.29. Let $\{\alpha_1, \ldots, \alpha_n\}$ be an integral basis of K and $d_K = \Delta(\alpha_1, \ldots, \alpha_n)$ the well-defined nonzero integer given by equation (3.3). We call d_K the *discriminant of K*.

3.4   Quadratic Fields

A number field K is a *quadratic field* if the degree $[K : \mathbb{Q}] = 2$. By Theorem 3.18, $K = \mathbb{Q}(\alpha)$, where α is a root of an irreducible polynomial $f(x) = ax^2 + bx + c$ of degree 2 over $\mathbb{Q}$. Since α is not a rational number, the discriminant $D = b^2 - 4ac$ of $f(x)$ cannot be zero or a perfect square. Write $D = dm^2$, with the integer $d \neq 0, 1$, having no square factor larger than 1. From the quadratic formula for solving quadratic polynomial equations, it is clear that $\mathbb{Q}(\alpha) = \mathbb{Q}(\sqrt{d})$. We summarize this as

Proposition 3.30. *A quadratic field $K = \mathbb{Q}(\sqrt{d})$ for a square-free integer $d \neq 0, 1$.*

The following theorem exhibits an integral basis of the ring of integers of a quadratic field explicitly.

Theorem 3.31. *Suppose $d \neq 0, 1$ is a square-free integer. Put*

$$
\omega = \begin{cases} \sqrt{d} & \text{if } d \equiv 2, 3 \pmod 4 \\ \frac{1+\sqrt{d}}{2} & \text{if } d \equiv 1 \pmod 4 \end{cases}
$$

Then $\{1, \omega\}$ is an integral basis of $K = \mathbb{Q}(\sqrt{d})$.

Proof. First we show that $\mathcal{O}_K \supseteq \mathbb{Z} + \mathbb{Z}\omega$. For this, all we need to show is that in case of $d \equiv 1 \pmod 4$, $\omega = (1 + \sqrt{d})/2$ is a root of a monic polynomial of degree 2 over $\mathbb{Z}$. It is easy to see that $x^2 - \operatorname{tr}_{K/\mathbb{Q}}(\omega)x + N_{K/k}(\omega) \in \mathbb{Z}[x]$ is such a polynomial. Next we show that $\mathcal{O}_K \subseteq \mathbb{Z} + \mathbb{Z}\omega$. Suppose $\alpha = a + b\sqrt{d} \in \mathcal{O}_K$ with $a, b \in \mathbb{Q}$. We know that $n = N_{K/k}(\alpha) = a^2 - db^2$, $m = \operatorname{tr}_{K/k}(\alpha) = 2a \in \mathbb{Z}$. Now if m is even, then $a \in \mathbb{Z} \Rightarrow db^2 \in \mathbb{Z}$. Since d is square-free, this implies that $b \in \mathbb{Z}$. This shows that $\alpha \in \mathbb{Z} + \mathbb{Z}\omega$. If m is odd, then $db^2 - \frac{1}{4} \in \mathbb{Z}$. Since d is square-free, $b = c/2$ with c odd. This gives $\omega = \frac{1+\sqrt{d}}{2}$ and $d \equiv 1 \pmod 4$.   □

Corollary 3.32. *The discriminant of the quadratic field $K = \mathbb{Q}(\sqrt{d})$, where $d \neq 0, 1$ is square-free, is given by*

$$
d_K = \begin{cases} 4d & \text{if } d \equiv 2, 3 \pmod 4 \\ d & \text{if } d \equiv 1 \pmod 4. \end{cases}
$$

Proof. The two $\mathbb{Q}$-homomorphisms $\sigma_i : K \to \mathbb{C}$ are the identity $\sigma_1 = 1_K$ and the conjugation σ_2 defined by $\sigma_2(x + y\sqrt{d}) = x - y\sqrt{d}$. Let $\{\alpha_1, \alpha_2\}$ be the integral basis of K given by Theorem 3.31. Using $d_K = (\det(\sigma_i(\alpha_j)))^2$, a short calculation is all we need.   □

Proposition 3.30 above can now be made more precise.

Theorem 3.33. *If K is a quadratic field, then $K = \mathbb{Q}(\sqrt{d_K})$, where d_K is the discriminant of K.*

3.5   Unique Factorization Property for Ideals

Let A be a commutative ring with 1. We recall the definition of an ideal of A. Suppose $\mathfrak{a}$ is a nonempty subset of A. We say that $\mathfrak{a}$ is an *ideal* of A, if for every $a \in A$ and $x, y \in \mathfrak{a}$, ax and $x + y \in \mathfrak{a}$. Every ideal contains 0, the zero element of A. The whole ring A itself is an ideal. An ideal $\mathfrak{a}$ of A is a *proper ideal* if $A \supsetneq \mathfrak{a}$. If $\mathfrak{a} \supsetneq \{0\}$, then we call $\mathfrak{a}$ a *nonzero ideal*.

Theorem 3.34. *If $\mathfrak{A}$ is a nonzero ideal of $\mathcal{O}_K$, then $\mathfrak{a} = \mathfrak{A} \cap \mathbb{Z}$ is a nonzero ideal of $\mathbb{Z}$.*

Proof. If $0 \neq \alpha \in \mathfrak{A}$, then α satisfies a nonzero monic polynomial over $\mathbb{Z}$, i.e.

$$a_0 + a_1 \alpha + \cdots + \alpha^n = 0,$$

with a_j in $\mathbb{Z}$ and $a_0 \neq 0$. Using the defining properties of an ideal, we see that $a_0 = -a_1 \alpha - \cdots - a_n \alpha^n \in \mathfrak{A} \cap \mathbb{Z} = \mathfrak{a}$. $\qquad\square$

Let $\mathfrak{a}$ be an ideal. The relation $x \sim y \Leftrightarrow x - y \in \mathfrak{a}$ is an equivalence relation which partitions A into disjoint sets of the form $x + \mathfrak{a} = \{x + a | a \in \mathfrak{a}\}$, called the *cosets* of $\mathfrak{a}$ in A. This set of cosets is a ring, called the *quotient* of A by $\mathfrak{a}$ and is denoted by $A/\mathfrak{a}$. The ring operations on $A/\mathfrak{a}$ are defined in an obvious way, namely,

$$(x + \mathfrak{a}) + (y + \mathfrak{a}) = (x + y) + \mathfrak{a}, \quad (x + \mathfrak{a})(y + \mathfrak{a}) = xy + \mathfrak{a}.$$

Remark 3.35. Let $\mathfrak{a}$ be an ideal of A. The notation $x \equiv y \pmod{\mathfrak{a}}$ means that $x - y \in \mathfrak{a}$.

Definition 3.36. Suppose $\mathfrak{m}$ is a proper ideal of A. We call $\mathfrak{m}$ a *maximal ideal* if for no other proper ideal $\mathfrak{a}$, we can have $\mathfrak{m} \subsetneq \mathfrak{a}$. We call a proper ideal $\mathfrak{p}$ a *prime ideal*, if, $a, b \in A$, $ab \in \mathfrak{p}$ implies that either $a \in \mathfrak{p}$ or $b \in \mathfrak{p}$.

Theorem 3.37. *Suppose $\mathfrak{a}$ is an ideal of A. Then*

1. $\mathfrak{a}$ is maximal if and only if $A/\mathfrak{a}$ is a field.

2. $\mathfrak{a}$ is prime if and only if $A/\mathfrak{a}$ is an integral domain.

Thus every maximal ideal is a prime ideal.

Arithmetic operations on ideals

Suppose $\mathfrak{a}, \mathfrak{b}$ are two ideals of a ring A. We define their sum and product as follows.

1. $\mathfrak{a} + \mathfrak{b} = \{a + b \,|\, a \in \mathfrak{a}, b \in \mathfrak{b}\}$,

2. $\mathfrak{a}\mathfrak{b}$ is the set of all finite sums $a_1 b_1 + \cdots + a_n b_n$ with $a_j \in \mathfrak{a}, b_j \in \mathfrak{b}$. It is the smallest ideal containing all the elements of the form ab with $a \in \mathfrak{a}, b \in \mathfrak{b}$.

Definition 3.38. Let $\mathfrak{a}$ and $\mathfrak{b}$ be two ideals of A with $\mathfrak{a} \neq (0)$. We say that $\mathfrak{a}$ *divides* $\mathfrak{b}$ if $\mathfrak{a} \supseteq \mathfrak{b}$. We write it as $\mathfrak{a}|\mathfrak{b}$.

Remark 3.39. An ideal $\mathfrak{a}$ of a commutative ring A with 1 is a *principal ideal* if it is generated by one element, i.e. $\mathfrak{a} = (a) = aA = \{ax \,|\, x \in A\}$. Note that $A = (1)$. We call A a *principal ideal ring* if every ideal of A is principal. A classical example of a principal ideal ring is $\mathbb{Z}$, and the above definition of divisibility of ideals was suggested by this example. For a, b in $\mathbb{Z}$, $(a)|(b)$ if and only if $a|b$.

We now prove Dedekind's famous theorem, namely, that for a number field K, if $\mathfrak{a} \neq (0), (1)$ is an ideal of $\mathcal{O}_K$, then it factors uniquely as

$$\mathfrak{a} = \mathfrak{p}_1^{e_1} \ldots \mathfrak{p}_r^{e_r}, \tag{3.4}$$

into a product of powers of distinct primes $\mathfrak{p}_1, \ldots, \mathfrak{p}_r$, taken in some fixed order. For $K = \mathbb{Q}$, $\mathcal{O}_K = \mathbb{Z}$ and in $\mathbb{Z}$, every ideal is principal. So if we identify the ideal (a) with the integer $a \geq 1$, the equation (3.4) becomes

$$a = p_1^{e_1} \ldots p_r^{e_r},$$

which says that every integer larger than 1 is a unique product of powers of distinct primes, taken in a fixed order. This is the *fundamental theorem of arithmetic*, and equation (3.4) generalizes the fundamental theorem of arithmetic to the ring of integers of a number field.

$\mathbb{Z}$-bases for nonzero ideals

Let A be a commutative ring with 1. Recall that an ideal $\mathfrak{a}$ of A is *finitely generated*, if it is finitely generated as an A-module, i.e. if there are finitely many elements $a_1, \ldots, a_n$ in $\mathfrak{a}$ such that

$$\mathfrak{a} = (a_1) + \cdots + (a_n).$$

The elements $a_1, \ldots, a_n$ are *generators* of $\mathfrak{a}$.

Definition 3.40. The ring A is called *Noetherian* (after the great German mathematician Emmy Noether, 1882–1935) if every ideal of A is finitely generated.

Clearly, a principal ideal ring is Noetherian. The following theorem from commutative algebra provides an alternative way of defining a Noetherian ring.

Theorem 3.41. *Suppose A is a commutative ring with 1. The following are equivalent:*

(1) Every non-empty family of ideals of A has a maximal element,

(2) A is Noetherian,

(3) Every ascending chain

$$\mathfrak{a}_1 \subseteq \mathfrak{a}_2 \subseteq \mathfrak{a}_3 \subseteq \cdots$$

of ideals is eventually stationary, i.e. there is an m, such that $\mathfrak{a}_n = \mathfrak{a}_m$ for all $n \geq m$.

Proof. (1) $\Rightarrow$ (2): Let $\mathfrak{a}$ be an ideal of A and S be the family of finitely generated submodules of A contained in $\mathfrak{a}$. Clearly, S is not empty. [The zero ideal is there.] By (1), let $\mathfrak{b}$ be a maximal element of this family. Then $\mathfrak{b} = \mathfrak{a}$ for otherwise, choosing a in $\mathfrak{a}$ but not in $\mathfrak{b}$, the finitely generated ideal $\mathfrak{c} = \mathfrak{b} + (a)$ would contradict the maximality of $\mathfrak{b}$.

(2) $\Rightarrow$ (3): By (2), the ideal $\mathfrak{a} = \cup \mathfrak{a}_j$ is finitely generated and all its generators are in $\mathfrak{a}_m$ for a sufficiently large m. Then for all $n \geq m$, $\mathfrak{a}_n = \mathfrak{a}_m$.

(3) $\Rightarrow$ (1): Apply Zorn's Lemma. We leave the details as an exercise. $\square$

We now show that for a number field K, the ring $\mathcal{O}_K$ is Noetherian. Actually, we have a stronger result, namely, that every nonzero ideal of $\mathcal{O}_K$ is a free $\mathbb{Z}$-module of maximum possible rank, which is $[K : \mathbb{Q}]$. For this we need the following fact.

Proposition 3.42. *If M is a free $\mathbb{Z}$-module of rank n and N is a submodule of M, then N is a free $\mathbb{Z}$-module of rank $r \leq n$. Moreover, if $M = \mathbb{Z}\alpha_1 \oplus \ldots \oplus \mathbb{Z}\alpha_n$, we can write $N = \mathbb{Z}\beta_1 \oplus \ldots \oplus \mathbb{Z}\beta_r$ with $r \leq n$, where*

$$\beta_i = \sum_{j=i}^{n} a_{ij} \alpha_j \tag{3.5}$$

and $a_{ij} \in \mathbb{Z}$ with $a_{jj} \geq 0$.

Proof. We use induction on n. If $n = 0$, there is nothing to prove. So let $n \geq 1$ and assume the theorem to be true for all M of rank $\leq n - 1$. Let $M' = \mathbb{Z}\alpha_2 \oplus \cdots \oplus \mathbb{Z}\alpha_n$, $N' = N \cap M'$.

The set $E = \{a \in \mathbb{Z} \mid a\alpha_1 + a_2\alpha_2 + \cdots + a_n\alpha_n \in N$ for some $a_2, \ldots, a_n$ in $\mathbb{Z}\}$, being an ideal of $\mathbb{Z}$, is of the form $a_{11}\mathbb{Z}$ with $a_{11} \geq 0$. Choose $\beta_1 = a_{11}\alpha_1 + a_{12}\alpha_2 + \cdots + a_{1n}\alpha_n$ in N. Then $N' = \{\alpha - aa_{11}\alpha_1 \mid \alpha = aa_{11}\alpha_1 + \cdots \in N\}$ is a submodule of M'. Since M' is a module of rank $n - 1$, we can choose β_j $(j \geq 2)$ as in equation (3.5). The set $\{\beta_1, \ldots, \beta_r\}$ is then a required $\mathbb{Z}$-basis of N. $\qquad \square$

Theorem 3.43. *Any nonzero ideal $\mathfrak{a}$ of $\mathcal{O}_K$ is a free $\mathbb{Z}$-module of rank $[K : \mathbb{Q}]$. In particular, $\mathcal{O}_K$ is Noetherian.*

Proof. By Proposition 3.42 above, we can certainly write

$$\mathfrak{a} = \mathbb{Z}\beta_1 + \cdots + \mathbb{Z}\beta_r$$

with $r \leq n$. All we need to show is that $r = n$. Let

$$\mathcal{O}_K = \mathbb{Z}\alpha_1 \oplus \cdots \oplus \mathbb{Z}\alpha_n.$$

By Theorem 3.34, choose $0 \neq a \in \mathfrak{a} \cap \mathbb{Z}$. Then $a\alpha_1, \ldots, a\alpha_n \in \mathfrak{a}$, which shows that the $\mathbb{Z}$-rank of $\mathfrak{a}$ is $\geq n$. Hence $r = n$. $\qquad \square$

EXERCISE

Show that the transition matrix between two $\mathbb{Z}$-bases of a nonzero ideal of $\mathcal{O}_K$ is unimodular, i.e. it is in $GL(n, \mathbb{Z})$.

Theorem 3.44. *For a nonzero ideal $\mathfrak{a}$ of $\mathcal{O}_K$, the quotient $\mathcal{O}_K/\mathfrak{a}$ is finite.*

Proof. Choose $a \neq 0$ in $\mathbb{Z} \cap \mathfrak{a}$. Then $a\mathcal{O}_K \subseteq \mathfrak{a} \subseteq \mathcal{O}_K$. This gives a surjective map $\mathcal{O}_K/a\mathcal{O}_K \to \mathcal{O}_K/\mathfrak{a}$. This proves the theorem, because the cardinality $|\mathcal{O}_K/\mathfrak{a}| \leq |\mathcal{O}_K/a\mathcal{O}_K| = a^n$. $\qquad \square$

Definition 3.45. The *norm* of a nonzero ideal $\mathfrak{a}$ of $\mathcal{O}_K$ is the positive integer $N(\mathfrak{a}) = |\mathcal{O}_K/\mathfrak{a}|$. We put $N((0)) = 0$.

Remark 3.46.

1. $N(\mathcal{O}_K) = 1$.

2. If $\mathfrak{a}|\mathfrak{b}$, then $N(\mathfrak{a}) \leq N(\mathfrak{b})$.

Theorem 3.47. *Let $\mathcal{O}_K = \mathbb{Z}\alpha_1 \oplus \ldots \oplus \mathbb{Z}\alpha_n$ and for a nonzero ideal $\mathfrak{a}$ of $\mathcal{O}_K$, let $\mathfrak{a} = \mathbb{Z}\beta_1 \oplus \ldots \oplus \mathbb{Z}\beta_n$ with $\beta_i = \sum_{j=i}^n a_{ij}\alpha_j$ $(a_{ii} > 0)$. Then $N(\mathfrak{a}) = a_{11} \ldots a_{nn}$.*

Proof. One can show by induction on $n = [K : \mathbb{Q}]$ that the set $\{r_i\alpha_i \mid 0 \leq r_i < a_{ii}\}$ forms a complete system of coset representatives of $\mathcal{O}_K/\mathfrak{a}$. We leave the details as an exercise. $\qquad\square$

Dedekind domains

We now define a Dedekind domain and show that

1. The ring of integers of a number field is a Dedekind domain.

2. In a Dedekind domain, the unique factorization theorem for ideals holds.

Definition 3.48. Suppose A is a subring of an integral domain B with 1 such that $1 \in A$. The *integral closure of A in B* is the set of elements of B that satisfy a monic polynomial over A.

Imitating the proof of Theorem 3.7, one can show that the integral closure of A in B is a subring of B containing A.

Definition 3.49. Suppose A is an integral domain and K is its field of fractions. We say that A is *integrally closed* if its integral closure in K is A.

We shall show that if K is a number field, then $\mathcal{O}_K$ is integrally closed.

Definition 3.50. Suppose A is a commutative ring with 1. The *Krull dimension* of A is the supremum of n such that there exist an ascending chain

$$(0) = \mathfrak{p}_0 \subsetneq \mathfrak{p}_1 \subsetneq \ldots \subsetneq \mathfrak{p}_n$$

of prime ideals in A. It is denoted by $\dim(A)$.

An important fact in algebraic geometry, that we shall neither need nor prove, is the following.

Theorem 3.51. $\dim(A[x]) = \dim(A) + 1$.

Example 3.52.

1. If K is a field, $\dim(K) = 0$.

2. The Krull dimension $\dim(A)$ can be infinite.

3. Since every maximal ideal is prime, $\dim(A) = 1 \Leftrightarrow$ every prime ideal of A is maximal.

4. The examples of rings A with $\dim(A) > 1$ are provided by algebraic geometry. For example, if k is a field and $A = k[x_1, \ldots, x_n]$, then $\dim(A) = n$. For a more interesting example, consider a nonsingular hypersurface X in an n-dimensional space defined as the zeros of a single irreducible polynomial $f(x_1, \ldots, x_n) \in k[x_1, \ldots, x_n]$. Its coordinate ring $A = k[X] = k[x_1, \ldots, x_n]/(f)$ has Krull dimension $n - 1$, the dimension of the hypersurface X as a geometric object. The study of an algebraic variety X, i.e. the zeros of a set of polynomials, is essentially the same as the study of its coordinate ring A (defined in the same manner as for a hypersurface). The geometric notions can thus be defined purely in ring theoretic terms, invented originally to study the arithmetic of number fields. For example, the intuitive notion of the dimension of a variety, which is one for curves and two for surfaces, coincides with the Krull dimension of its coordinate ring. This leads one to consider algebraic geometry as an extension of algebraic number theory.

Definition 3.53. An integral domain A is a *Dedekind domain* if

1. A is Noetherian,

2. A is integrally closed and

3. $\dim(A) = 1$.

Theorem 3.54. *The ring $\mathcal{O}_K$, of integers of a number field K, is a Dedekind domain.*

We have already shown (Theorem 3.43) that for a number field K, its ring of integers $\mathcal{O}_K$ is Noetherian. The next two theorems show that it is a Dedekind domain.

Theorem 3.55. *The ring $\mathcal{O}_K$ is integrally closed.*

Proof. We need to show that if $\alpha \in K$ satisfies a monic polynomial over $\mathcal{O}_K$, then it is in $\mathcal{O}_K$. By Proposition 3.10, $\mathcal{O}_K[\alpha]$ is a finitely generated $\mathcal{O}_K$-module. Since $\mathcal{O}_K = \mathbb{Z}\alpha_1 + \cdots + \mathbb{Z}\alpha_n$, $\mathbb{Z}[\alpha]$ is a finitely generated $\mathbb{Z}$-module. By Proposition 3.10 again, α is an algebraic integer and hence is in $\mathcal{O}_K$. $\square$

Theorem 3.56. *The ring $\mathcal{O}_K$ has Krull dimension 1.*

Proof. First note that a finite domain $A = \{x_1, \ldots, x_m\}$ is a field. [To see this let $0 \neq x \in A$. Then $xx_1, \ldots, xx_m$ are all distinct, hence one of them has to be equal to 1.] Now if $\mathfrak{p}$ is a prime ideal of $\mathcal{O}_K$, then $\mathcal{O}_K/\mathfrak{p}$ is a finite domain, which shows that it is a field. But then $\mathfrak{p}$ has to be maximal. $\square$

Fractional ideals

For the rest of this section, we assume that A is a Dedekind domain and prove that its ideals have the unique factorization property. We denote by K the field of fractions of A.

Definition 3.57. A subset $\mathfrak{b}$ of K is a *fractional ideal* of K if $\mathfrak{b}$ is a nonzero A-module and for a nonzero $d \in A$, $d\mathfrak{b} \subseteq A$.

To avoid confusion, we shall call an ideal $\mathfrak{a} \subseteq A$ an *integral ideal*. An integral ideal is a fractional ideal. [Take $d = 1$.] If $\mathfrak{b}$ is a fractional ideal, then so is $d^{-1}\mathfrak{b}$ for every nonzero $d \in A$. If $\mathfrak{a}, \mathfrak{b}$ are fractional ideals, then so are $\mathfrak{a} + \mathfrak{b}$ and $\mathfrak{a}\mathfrak{b}$. Suppose $\mathfrak{b}$ is a fractional ideal. Choose d in A such that $\mathfrak{a} = d\mathfrak{b} \subseteq A$. Then $\mathfrak{a}$ is an (integral) ideal of A. Therefore, when $A = \mathcal{O}_K$, every nonzero fractional ideal is a free $\mathbb{Z}$-module of rank $n = [K : \mathbb{Q}]$. Finally, we prove the following theorem of Dedekind.

Theorem 3.58. *Any nonzero ideal $\mathfrak{a} \neq (1)$ of a Dedekind domain A is a unique product*

$$\mathfrak{a} = \mathfrak{p}_1^{e_1} \ldots \mathfrak{p}_r^{e_r}$$

of powers of distinct prime ideals, taken in some fixed order.

For the proof, we need following three lemmas from commutative algebra. Remember that A is a Dedekind domain.

Lemma 3.59. *If a prime ideal $\mathfrak{p}$ divides the product $\mathfrak{a}_1 \ldots \mathfrak{a}_r$ of ideals, then $\mathfrak{p}$ divides $\mathfrak{a}_j$ for some j.*

Recall that $\mathfrak{a}|\mathfrak{b}$ if and only if $\mathfrak{a} \supseteq \mathfrak{b}$.

Proof. If not, for each j we choose an a_j in $\mathfrak{a}_j$ but not in $\mathfrak{p}$. Then $a_1 \ldots a_r \in \mathfrak{a}_1 \ldots \mathfrak{a}_r \subseteq \mathfrak{p}$. Since $\mathfrak{p}$ is prime, some $a_j \in \mathfrak{p}$. This is a contradiction. $\qquad\square$

Lemma 3.60. *Suppose A is not a field. Then every nonzero ideal of A contains a product of prime ideals.*

Proof. Let S be the family of nonzero ideals that fail the theorem. We show that S is empty. If not, let $\mathfrak{a}$ be a maximal element of this family, which exists because A is Noetherian. Clearly, $\mathfrak{a} \neq A$ since, being a domain, A certainly contains the prime ideal (0). Also $\mathfrak{a}$ is not prime, which implies that there are a_1, a_2 in A but not in $\mathfrak{a}$, such that $a_1 a_2 \in \mathfrak{a}$. Put $\mathfrak{a}_j = \mathfrak{a} + (a_j)$. Then $\mathfrak{a}_j \supsetneq \mathfrak{a}$. Hence, $\mathfrak{a}_1 \supseteq \mathfrak{p}_1 \ldots \mathfrak{p}_r$ and $\mathfrak{a}_2 \supseteq \mathfrak{q}_1 \ldots \mathfrak{q}_s$. This implies that $\mathfrak{a} \supseteq \mathfrak{a}_1 \mathfrak{a}_2 \supseteq \mathfrak{p}_1 \ldots \mathfrak{p}_r \mathfrak{q}_1 \ldots \mathfrak{q}_s$. This is a contradiction. $\qquad\square$

Lemma 3.61. *In a Dedekind domain A, every nonzero prime ideal $\mathfrak{p}$ is invertible, i.e. $\mathfrak{p}\mathfrak{q} = (1) = A$, for some fractional ideal $\mathfrak{q}$.*

We call $\mathfrak{q}$ the *inverse ideal* of $\mathfrak{p}$ and denote it by $\mathfrak{p}^{-1}$.

Proof. Let $\mathfrak{p}^{-1} = \{d \in K | d\mathfrak{p} \subseteq A\}$. Clearly, $\mathfrak{p}^{-1}$ is an A-module containing A and $\mathfrak{p} \subseteq \mathfrak{p}^{-1}\mathfrak{p} \subseteq A$. Since $\mathfrak{p}$ is maximal, either $\mathfrak{p}^{-1}\mathfrak{p} = A$, in which case we are done, or $\mathfrak{p}^{-1}\mathfrak{p} = \mathfrak{p}$, which we show is impossible. Suppose $\mathfrak{p}^{-1}\mathfrak{p} = \mathfrak{p}$. Then for each $x \in \mathfrak{p}^{-1}$, $x\mathfrak{p} \subseteq \mathfrak{p}$, $x^2\mathfrak{p} \subseteq \mathfrak{p}, \ldots, x^n\mathfrak{p} \subseteq \mathfrak{p}, \ldots$ Taking $M = \mathfrak{p}$ in Proposition 3.10, we see that x is integral over A. Since A is integrally closed, $x \in A$ which implies that $\mathfrak{p}^{-1} = A$. We show that this is impossible.

Suppose $\mathfrak{p}^{-1} = A$. Choose $0 \neq a \in \mathfrak{p}$. Then by Lemma 3.60, $\mathfrak{p} \supseteq aA \supseteq \mathfrak{p}_1 \ldots \mathfrak{p}_r$. Choose r the smallest possible. By Lemma 3.59, (after renumbering) $\mathfrak{p} \supseteq \mathfrak{p}_1$. Since $\dim(A) = 1$, $\mathfrak{p} = \mathfrak{p}_1$. Put $\mathfrak{a} = \mathfrak{p}_2 \ldots \mathfrak{p}_r$. [If $r = 1$, then $\mathfrak{a} = A$.] By minimality of r, $aA \not\supseteq \mathfrak{a}$. So choose $x \in \mathfrak{a}, x \notin aA$. Then $\mathfrak{p}\mathfrak{a} \subseteq aA \Rightarrow x\mathfrak{p} \subseteq aA \Rightarrow a^{-1}x\mathfrak{p} \subseteq A \Rightarrow a^{-1}x \in \mathfrak{p}^{-1} = A \Rightarrow x \in aA$, which is a contradiction. $\square$

Proof of Dedekind's Theorem

Existence. Suppose S is the set of ideals of A, $\neq (0)$, (1), for which the theorem is false. We show that S is empty. If not, choose a maximal element $\mathfrak{a}$ in S. Then $\mathfrak{a}$ is not prime and hence $\mathfrak{a} \subsetneq \mathfrak{p}$, for a prime ideal $\mathfrak{p}$. Since $A \subseteq \mathfrak{p}^{-1}$, we have $\mathfrak{a} \subseteq \mathfrak{a}\mathfrak{p}^{-1} \subsetneq \mathfrak{p}\mathfrak{p}^{-1} = A$. First we show that $\mathfrak{a} \subsetneq \mathfrak{a}\mathfrak{p}^{-1}$. By taking $M = \mathfrak{a}$ in Proposition 3.10, the equality $\mathfrak{a}\mathfrak{p}^{-1} = \mathfrak{a}$ implies that every $x \in \mathfrak{p}^{-1}$ is integral over A. Since A is integrally closed, $x \in A$, which shows that $\mathfrak{p}^{-1} = A$, i.e. $\mathfrak{p} = A$, which is not the case. Now by maximality of $\mathfrak{a}$, we have $\mathfrak{a}\mathfrak{p}^{-1} = \mathfrak{p}_2 \ldots \mathfrak{p}_r \Rightarrow \mathfrak{a} = \mathfrak{p}\mathfrak{p}_2 \ldots \mathfrak{p}_r$. This contradicts that $\mathfrak{a}$ is in S.

Uniqueness. If $\mathfrak{a} = \mathfrak{p}_1 \ldots \mathfrak{p}_r = \mathfrak{q}_1 \ldots \mathfrak{q}_s$ has two factorizations, we have by Lemma 3.59, $\mathfrak{p}_1 \supseteq \mathfrak{q}_1$. But since $\dim(A) = 1$, $\mathfrak{p}_1 = \mathfrak{q}_1$. By Lemma 3.61, we can cancel $\mathfrak{p}_1$ to get $\mathfrak{p}_2 \ldots \mathfrak{p}_r = \mathfrak{q}_2 \ldots \mathfrak{q}_s$ and so on. This process must end with no prime ideal left on either side. $\square$

Remark 3.62. It is clear from the proof of Dedekind's theorem that the inverse of a prime ideal is a fractional ideal. Hence the nonzero fractional ideals of K (K being the quotient field of a Dedekind domain A) form an Abelian group under the multiplication of ideals. A is the identity of this group.

Corollary 3.63. *If $\mathfrak{a}$ is a fractional ideal of a Dedekind domain, then except for ordering the prime ideals, we have a unique factorization*

$$\mathfrak{a} = \frac{\mathfrak{p}_1 \ldots \mathfrak{p}_r}{\mathfrak{q}_1 \ldots \mathfrak{q}_s}.$$

Here $\frac{1}{\mathfrak{q}}$ means $\mathfrak{q}^{-1}$. For any nonzero ideal $\mathfrak{a}$ of A, we set $\mathfrak{a}^0 = (1) = A$.

Let $\mathfrak{a}$ be a fractional ideal of a Dedekind domain A for a nonzero prime ideal $\mathfrak{p}$ of A, the *discrete valuation* $v_\mathfrak{p}(\mathfrak{a})$ *of* $\mathfrak{a}$ *at a prime ideal* $\mathfrak{p}$ is defined to be the exponent to which it appears in the unique factorization of $\mathfrak{a}$ into powers of distinct prime ideals. It is an integer (positive, negative or zero).

For the map $v_{\mathfrak{p}} : K \to \mathbb{Z}$ given by $v_{\mathfrak{p}}(\alpha) = v_{\mathfrak{p}}((\alpha))$, $\alpha \neq 0$ and $v_{\mathfrak{p}}(0) = \infty$, we have

1. $v_{\mathfrak{p}}(\alpha\beta) = v_{\mathfrak{p}}(\alpha) + v_{\mathfrak{p}}(\beta)$, and

2. if $v_{\mathfrak{p}}(\alpha) \neq v_{\mathfrak{p}}(\beta)$, then $v_{\mathfrak{p}}(\alpha + \beta) = \min(v_{\mathfrak{p}}(\alpha), v_{\mathfrak{p}}(\beta))$.

Definition 3.64. A map $v : K^{\times} \to \mathbb{Z}$ with properties 1 and 2 is called a *discrete valuation* on K.

Two discrete valuations on K are equivalent, if they can be scaled to give the same valuation. Among all the equivalent valuations, there is a unique one which subjects to $\mathbb{Z}$. We will use this normalized valuation to represent all the valuations equivalent to it. For each prime $\mathfrak{p}$ of $\mathcal{O}_K$, $v_{\mathfrak{p}}$ is a normalized discrete valuation on K.

Definition 3.65. The *greatest common divisor* (briefly, g.c.d.) of two ideals $\mathfrak{a}, \mathfrak{b}$ is the ideal

$$(\mathfrak{a}, \mathfrak{b}) = \prod_{\mathfrak{p}} \mathfrak{p}^{\min(v_{\mathfrak{p}}(\mathfrak{a}), v_{\mathfrak{p}}(\mathfrak{b}))}.$$

Definition 3.66. Two ideals $\mathfrak{a}$ and $\mathfrak{b}$ are *coprime* if the g.c.d. $(\mathfrak{a}, \mathfrak{b}) = (1)$.

The *least common multiple* $[\mathfrak{a}, \mathfrak{b}]$ is defined by replacing min with max in the definition of the greatest common divisor.

Let $\mathfrak{a}$ and $\mathfrak{b}$ be two ideals of a Dedekind domain A. To say that $\mathfrak{a}$ *divides* $\mathfrak{b}$ is equivalent to saying that $v_{\mathfrak{p}}(\mathfrak{a}) \leq v_{\mathfrak{p}}(\mathfrak{b})$ for all prime ideals $\mathfrak{p}$.

EXERCISES

1. Show that $v_{\mathfrak{p}}$ is a normalized discrete valuation on a number field K.

2. If $\mathfrak{a}, \mathfrak{b}$ are nonzero ideals of K, show that the g.c.d. $(\mathfrak{a}, \mathfrak{b}) = \mathfrak{a} + \mathfrak{b}$. Thus $\mathfrak{a}$ and $\mathfrak{b}$ are coprime $\Leftrightarrow \mathfrak{a} + \mathfrak{b} = (1) = A$.

3. Show that a principal ideal domain is a Dedeknd domain.

4. Show that $\mathbb{Z}[\sqrt{-5}]$ is a Dedekind domain, but not a PID.

3.6   Ideal Class Group and Class Number

In the last section, we proved that the nonzero fractional ideals of a Dedekind domain A with quotient field K, form an Abelian group I under the operation of multiplication of ideals. The nonzero principal fractional ideals, that is the ideals of the form $\alpha A = \{\alpha a | a \in A\}$ with $\alpha \neq 0$ in K, form a subgroup P of I. The quotient group I/P is called the *ideal class group* of K. The elements of I/P are called the *ideal classes*. The cardinality of the ideal class group is called the *class number* of K. We will denote the class number of K by h_K. In this section we shall show that the class number of a number field is finite, in which $A = \mathcal{O}_K$.

Recall that for a nonzero ideal $\mathfrak{a}$ of $\mathcal{O}_K$, its norm $N(\mathfrak{a})$ is the cardinality of the quotient ring $\mathcal{O}_K/\mathfrak{a}$. We have seen that this cardinality is finite.

Theorem 3.67. *Suppose* $(\alpha) = \alpha\mathcal{O}_K$ *is a principal ideal of* $\mathcal{O}_K$. *Then we have*

$$N((\alpha)) = |N_{K/\mathbb{Q}}(\alpha)|.$$

Proof. If $\alpha = 0$ there is nothing to prove, Otherwise, write $\mathcal{O}_K = \mathbb{Z}\alpha_1 \oplus \ldots \oplus \mathbb{Z}\alpha_n$, where $n = [K : \mathbb{Q}]$. By Proposition 3.42, we can also write $((\alpha)) = \mathbb{Z}\beta_1 \oplus \ldots \oplus \mathbb{Z}\beta_n$, where

$$\beta_i = \sum_{j=i}^{n} a_{ji}\alpha_j$$

with $a_{ii} > 0$. By Theorem 3.47, $N((\alpha)) = a_{11} \ldots a_{nn}$. On the other hand,

$$(\alpha) = \mathbb{Z}\alpha\alpha_1 \oplus \cdots \oplus \mathbb{Z}\alpha\alpha_n$$

which shows that $\{\alpha\alpha_1, \ldots, \alpha\alpha_n\}$ is a $\mathbb{Z}$-basis of (α). The transition matrix U from $\{\beta_1, \ldots, \beta_n\}$ to $\{\alpha\alpha_1, \ldots, \alpha\alpha_n\}$ is unimodular. If for $i < j$ we let $a_{ij} = 0$ and put $M = (a_{ij})$, then

$$\alpha \begin{pmatrix} \alpha_1 \\ \vdots \\ \alpha_n \end{pmatrix} = UM \begin{pmatrix} \alpha_1 \\ \vdots \\ \alpha_n \end{pmatrix}.$$

Therefore,

$$|N_{K/\mathbb{Q}}(\alpha)| = |\det(UM)| = |\det(M)| = |N(\alpha\mathcal{O}_K)|. \qquad \square$$

Corollary 3.68. *If* $a \in \mathbb{N}$, *then* $N(a\mathcal{O}_K) = a^n$.

Theorem 3.69 (Independence of Valuations). *Given distinct prime ideals* $\mathfrak{p}_1, \ldots, \mathfrak{p}_r$ *of a Dedekind domain* A *and integers* $a_1, \ldots, a_r > 0$, *there is* α *in* A, *such that* $v_{\mathfrak{p}_i}(\alpha) = a_i$, *for all* i.

Proof. Choose α_i in $\mathfrak{p}_1^{a_1} \ldots \mathfrak{p}_r^{a_r}$ which is not in $\mathfrak{p}_1^{a_1} \ldots \mathfrak{p}_{i-1}^{a_{i-1}}\mathfrak{p}_i^{a_i+1}\mathfrak{p}_{i+1}^{a_{i+1}} \ldots \mathfrak{p}_r^{a_r}$ and put $\alpha = \alpha_1 + \cdots + \alpha_r$. $\qquad \square$

Theorem 3.70. *There is an α in K with prescribed "zeros" and "poles", i.e. given distinct prime ideals $\mathfrak{p}_1, \ldots, \mathfrak{p}_r; \mathfrak{q}_1, \ldots, \mathfrak{q}_s$ of a Dedekind domain A and positive integers $a_1, \ldots, a_r; b_1, \ldots, b_s$, there is an α in the quotient field K of A, such that $v_{\mathfrak{p}_i}(\alpha) = a_i$ and $v_{\mathfrak{q}_j}(\alpha) = -b_j$.*

Proof. Choose β and γ in A with $v_{\mathfrak{p}_i}(\beta) = a_i$, $v_{\mathfrak{q}_j}(\gamma) = b_j$ and put $\alpha = \beta/\gamma$. □

Corollary 3.71. *For two ideals $\mathfrak{a}$ and $\mathfrak{b}$ of a Dedekind domain A, there is an α in A with the g.c.d. $(\mathfrak{ab}, (\alpha)) = \mathfrak{a}$.*

Proof. If $\mathfrak{a} = \mathfrak{p}_1^{a_1} \ldots \mathfrak{p}_r^{a_r}$, choose α as in Theorem 3.69. □

Theorem 3.72. *Any ideal $\mathfrak{a}$ of a Dedekind domain A can be generated by two elements.*

Proof. It is easy to see that given two ideals $\mathfrak{a}$ and $\mathfrak{b}$ of A, the g.c.d. $(\mathfrak{a}, \mathfrak{b}) = \mathfrak{a} + \mathfrak{b}$. Choose β in K, such that $\beta\mathfrak{a}^{-1} = \mathfrak{b}$ is an integral ideal of A. Then $\mathfrak{ab} = (\beta)$. Now choose α, such that $\mathfrak{a} = (\mathfrak{ab}, (\alpha)) = ((\beta), (\alpha)) = \beta A + \alpha A$. □

Corollary 3.73. *For any nonzero ideal $\mathfrak{a}$ of a Dedekind domain A, the quotient ring $A/\mathfrak{a}$ is a principal ideal domain.*

Theorem 3.74. *The norm is multiplicative, i.e. if $\mathfrak{a}$ and $\mathfrak{b}$ are two integral ideals of $\mathcal{O}_K$, then*

$$N(\mathfrak{ab}) = N(\mathfrak{a})N(\mathfrak{b}).$$

Proof. Let $N(\mathfrak{a}) = r$ and $N(\mathfrak{b}) = s$. Choose coset representatives $\{\alpha_1, \ldots, \alpha_r\}$ and $\{\beta_1, \ldots, \beta_s\}$ of $\mathcal{O}_K/\mathfrak{a}$ and $\mathcal{O}_K/\mathfrak{b}$, respectively and choose α as in Corollary 3.71. We show that $\{\alpha_i + \alpha\beta_j | i = 1, \ldots, r; j = 1, \ldots, s\}$ is a *"complete set of coset representatives"* of $\mathcal{O}_K/\mathfrak{ab}$ (which means the following).

1. *They are distinct* mod $\mathfrak{ab}$. If $\alpha_i + \alpha\beta_j \equiv \alpha_l + \alpha\beta_m$ $(mod\ \mathfrak{ab})$, then $(\alpha_i - \alpha_l) + \alpha(\beta_j - \beta_m) \in \mathfrak{ab} \subseteq \mathfrak{a}$. But $\alpha \in \mathfrak{a}$. Therefore, $\alpha_i - \alpha_l \in \mathfrak{a} \Rightarrow i = l \Rightarrow \alpha(\beta_j - \beta_m) \in \mathfrak{ab}$. However, by our choice of α, this implies that $\beta_j - \beta_m$ is in $\mathfrak{b}$, which gives $j = m$.

2. *The set contains a representative of every coset of $A/\mathfrak{ab}$.* If x is in $\mathcal{O}_K$, we can write $x = \alpha_i + a$ for some a in $\mathfrak{a}$. But $(\mathfrak{ab}, (\alpha)) = \mathfrak{a}$. Therefore, $a = \beta\alpha + c$, $c \in \mathfrak{ab}$. Now $\beta = \beta_j + b$, $b \in \mathfrak{b}$ for some j. Hence $x = \alpha_i + \alpha\beta_j + y$ for some y in $\mathfrak{ab}$. □

EXERCISE

Let $\mathfrak{P}$ be a prime ideal of $\mathcal{O}_K$. Show that $\mathfrak{P} \cap \mathbb{Z} = \mathfrak{p}$ is a prime ideal of $\mathbb{Z}$. [Hence, $\mathfrak{p} = p\mathbb{Z}$ for a unique prime number p.]

Theorem 3.75. *There are only finitely many ideals of norm less than a given constant c.*

Proof. By the multiplicative property of the norm, it is enough to prove the theorem for prime ideals $\mathfrak{P}$ of $\mathcal{O}_K$. Given $\mathfrak{P}$, there is a unique prime p in $\mathbb{Z}$, such that $\mathfrak{P} \cap \mathbb{Z} = p\mathbb{Z}$. Then $\mathfrak{P}$ divides $(p) = p\mathcal{O}_K$. Therefore,

$$N(\mathfrak{P}) \leq |N((p))| = p^n,$$

where $n = [K : k]$. But given a constant $c > 0$, there are only finitely many primes p with $p \leq c$, and given such a prime p, there are only finitely many $\mathfrak{P}$ that can divide (p). Only primes $\mathfrak{P}$ that divide a $p < c$ can have norm less than c. $\qquad\square$

Theorem 3.76. *There is a constant $c = c(K)$, such that each nonzero ideal $\mathfrak{a} \subseteq \mathcal{O}_K$ contains an element $\alpha \neq 0$, with $|N(\alpha)| \leq cN(\mathfrak{a})$.*

Proof. We use the pigeon-hole principle. Write

$$\mathcal{O}_K = \mathbb{Z}\alpha_1 \oplus \cdots \oplus \mathbb{Z}\alpha_n$$

Let t be the smallest integer $\geq N(\mathfrak{a})^{1/n}$ and look at the $(t+1)^n$ distinct elements

$$a_1\alpha_1 + \cdots + a_n\alpha_n$$

of $\mathcal{O}_K$, where the integers a_j satisfy $0 \leq a_j \leq t$. Since $N(\mathfrak{a}) < (t+1)^n$, among them there are two distinct ones which represent the same element of $\mathcal{O}_K/\mathfrak{a}$. We show that their difference $\alpha = c_1\alpha_1 + \cdots + c_n\alpha_n$ is the required nonzero element of $\mathfrak{a}$. Let L_α be the matrix of the linear map "multiplication by α" with respect to the basis $\{\alpha_1, \ldots, \alpha_n\}$ of $K/\mathbb{Q}$. Similarly, let L_j be the matrix of α_j. By linearity of the map $\alpha \to L_\alpha$, we have $L_\alpha = c_1L_1 + \cdots + c_nL_n$. Since $|c_j| \leq t$, we have

$$|N(\alpha)| = |\det(L_\alpha)| = |\det(c_1L_1 + \cdots + c_nL_n)| \leq c(\alpha_1, \ldots, \alpha_n) \max |c_j|^n$$
$$\leq ct^n \leq cN(\mathfrak{a}),$$

where the constant $c = c(\alpha_1, \ldots, \alpha_n)$ depends only on $\alpha_1, \ldots, \alpha_n$ and hence only on K. $\qquad\square$

EXERCISE

Let $K = \mathbb{Q}(\sqrt{d})$ be a quadratic field with square-free $d \neq 0, 1$. Calculate $c(\alpha_1, \ldots, \alpha_n)$.

Theorem 3.77 (Dedekind)**.** *The class number h_K of a number field K is finite.*

Proof. If $\mathcal{K}$ is an ideal class, choose an integral ideal $\mathfrak{a}$ in $\mathcal{K}^{-1}$ and a nonzero α in $\mathfrak{a}$ with $|N(\alpha)| \leq cN(\mathfrak{a})$. Put $\mathfrak{b} = (\alpha)\mathfrak{a}^{-1}$, so that $\mathfrak{a}\mathfrak{b} = (\alpha)$. Then

$$N(\mathfrak{a})N(\mathfrak{b}) = N(\mathfrak{a}\mathfrak{b}) = N((\alpha)) = |N(\alpha)| \leq cN(\mathfrak{a}).$$

This gives $N(\mathfrak{b}) \leq c$ which implies that $\mathcal{K}$ contains an ideal $\mathfrak{b}$ of norm $\leq c$. Since there are only finitely many ideals of bounded norm, there are only finitely many ideal classes $\mathcal{K}$. $\square$

4

Arithmetic in Relative Extensions

Throughout this chapter, K will denote a number field and k a subfield of K. The extension K/k will be called a *relative extension of number fields*. We put $\mathfrak{o} = \mathcal{O}_k$ and $\mathcal{O} = \mathcal{O}_K$.

Theorem 4.1. $\mathcal{O}_K = \{\alpha \in K \mid f(\alpha) = 0 \text{ for a monic polynomial } f(x) \text{ in } \mathfrak{o}[x]\}$

Proof. We only have to show that any α in K which satisfies a monic polynomial $f(x)$ in $\mathfrak{o}[x]$ is an algebraic integer. Let

$$f(x) = a_0 + a_1 x + \cdots + a_{n-1} x^{n-1} + x^n$$

with a_j in $\mathfrak{o}$. Since a_j are algebraic integers,

$$M = \mathbb{Z}[a_0, a_1, \ldots, a_{n-1}]$$

is a finitely generated $\mathbb{Z}$-module, and so is

$$\mathbb{Z}[a_0, a_1, \ldots, a_{n-1}, \alpha] = M + M\alpha + \cdots + M\alpha^{n-1}.$$

Since $\mathbb{Z}[\alpha]$ is a submodule of a finitely generated $\mathbb{Z}$-module, $\mathbb{Z}[\alpha]$ is also a finitely generated $\mathbb{Z}$-module, which shows that α is an algebraic integer. $\square$

Remark 4.2. This theorem allows us to regard $K/\mathbb{Q}$ as a special case of the relative extension K/k of number fields with $k = \mathbb{Q}$.

EXERCISE

Let $k \subseteq K \subseteq L$ be number fields with $[K : k] = n$ and $[L : K] = m$. Let $\sigma_1, \ldots, \sigma_n$ be the distinct k-isomorphisms of K into $\mathbb{C}$. Show that each σ_i extends to m distinct k-isomorphisms $\sigma_{ij} : L \to \mathbb{C}$.

Hint: Let $L = K(\theta)$ and $\tau_1, \ldots, \tau_m$ be the m distinct K-isomorphisms of L into $\mathbb{C}$. If we write α in L as

$$\alpha = a_0 + a_1 \theta + \cdots + a_{m-1} \theta^{m-1}$$

with coefficients in K, put

$$\sigma_{ij}(\alpha) = \sigma_i(a_0) + \sigma_i(a_1)\tau_j(\theta) + \cdots + \sigma_i(a_{m-1})\tau_j(\theta^{m-1}).$$

Recall the following.

Definition 4.3. For a relative extension K/k of number fields of degree n, the *relative norm*

$$N_{K/k} : K \to k$$

is given by

$$N(\alpha) = \sigma_1(\alpha) \cdots \sigma_n(\alpha),$$

where $\sigma_1, \ldots, \sigma_n : K \to \mathbb{C}$ are all the k-isomorphisms of K into $\mathbb{C}$.

The *relative trace*

$$\mathrm{Tr}_{K/k} : K \to k$$

is defined similarly by

$$\mathrm{Tr}_{K/k}(\alpha) = \sigma_1(\alpha) + \cdots + \sigma_n(\alpha).$$

Clearly,

1. $N_{K/k}(\alpha\beta) = N_{K/k}(\alpha)N_{K/k}(\beta)$ and $N_{K/k}(a) = a^n$, if $a \in k$.

2. $\mathrm{Tr}_{K/k}(\alpha + \beta) = \mathrm{Tr}_{K/k}(\alpha) + \mathrm{Tr}_{K/k}(\beta)$ and $\mathrm{Tr}_{K/k}(a) = na$, if $a \in k$.

EXERCISES

1. Suppose $k \subseteq K \subseteq L$. Show that

   (a) $N_{L/k} = N_{K/k} \cdot N_{L/K}$, and

   (b) $\mathrm{Tr}_{L/k} = \mathrm{Tr}_{K/k} \cdot \mathrm{Tr}_{L/K}$.

   [*Hint*: Use Exercise 4.]

2. Let σ be a k-isomorphism of K into $\mathbb{C}$ for the relative extension K/k of number fields. If $\mathfrak{A}$ is an ideal in $\mathcal{O}$, show that $\sigma(\mathfrak{A})$ is an ideal in the ring $\mathcal{O}_{\sigma(K)}$ of integers of $\sigma(K)$.

3. Let K/k be an extension of number fields and L the normal closure of K in $\mathbb{C}$. [Recall that L is the smallest subfield of $\mathbb{C}$ containing $\sigma_i(K)$ for all the k-isomorphisms $\sigma_1, \ldots, \sigma_n$ of K into $\mathbb{C}$.] Let $\mathfrak{A}$ be an ideal in $\mathcal{O} = \mathcal{O}_K$. For each j, $\sigma_j(\mathfrak{A})$ is an ideal of $\sigma_j(\mathcal{O})$. Let $\sigma_1(\mathfrak{A}) \cdots \sigma_n(\mathfrak{A})$ denote the ideal of $\mathcal{O}_L$ generated by the products $\alpha_1 \cdots \alpha_n$ with α_j in $\sigma_j(\mathfrak{A})$. We call the ideal $\sigma_1(\mathfrak{A}) \cdots \sigma_n(\mathfrak{A}) \cap \mathfrak{o}$ of $\mathfrak{o}$ the *relative norm of the ideal* $\mathfrak{A}$ and denote it by $N_{K/k}(\mathfrak{A})$.

   Show that

   i) $N_{K/k}(\mathfrak{A}\mathfrak{B}) = N_{K/k}(\mathfrak{A})N_{K/k}(\mathfrak{B})$, and

   ii) if $\mathfrak{a}$ is an ideal in $\mathfrak{o}$, then $N_{K/k}(\mathfrak{a}) = \mathfrak{a}^n$.

Now suppose that K/k is an arbitrary but fixed extension of number fields. Let $\mathfrak{o} = \mathcal{O}_k$ and $\mathcal{O} = \mathcal{O}_K$. We know that any nonzero ideal in $\mathfrak{o}$ is a unique product of powers of prime ideals in $\mathfrak{o}$ and the same is true for $\mathcal{O}$ also. However, a prime ideal $\mathfrak{p}$ of $\mathfrak{o}$ need not generate a prime ideal of $\mathcal{O}$.

Example 4.4. If $k = \mathbb{Q}$, $K = \mathbb{Q}(i)$, then for $\alpha = 2 + i$,

$$5\mathcal{O} = (\alpha)(\overline{\alpha}).$$

Since $N(\alpha) = N(\overline{\alpha}) = 5$ is not a unit in $\mathbb{Z}$, (α) and $(\overline{\alpha})$ are both proper ideals, each is divisible by a prime ideal of $\mathcal{O}$. Hence, $5\mathcal{O}$ is not a prime ideal of $\mathcal{O}$.

Suppose $\mathfrak{p}$ is a prime ideal in $\mathfrak{o}$ and $\mathfrak{p}\mathcal{O}$ the ideal generated by $\mathfrak{p}$ in $\mathcal{O}$. Let

$$\mathfrak{p}\mathcal{O} = \mathfrak{P}_1^{e_1} \cdots \mathfrak{P}_g^{e_g} \qquad (e_g \geq 1) \tag{4.1}$$

be its unique factorization into powers of distinct primes $\mathfrak{P}_1, \ldots, \mathfrak{P}_g$ in $\mathcal{O}$. We call $\mathfrak{P}_1, \ldots, \mathfrak{P}_g$ the *prime divisors* of $\mathfrak{p}$ in $\mathcal{O}_K$, or by abuse of language, in K. The positive integers $e_j = e(\mathfrak{P}_j/\mathfrak{p})$ are the *exponents* of $\mathfrak{P}_j$ over $\mathfrak{p}$.

If $\mathfrak{P}$ is a prime divisor of $\mathfrak{p}$ in K, we write it as $\mathfrak{P}|\mathfrak{p}$. If $\mathfrak{P}|\mathfrak{p}$, then $\mathfrak{P} \cap \mathfrak{o} = \mathfrak{p}$.

Definition 4.5. Let $\mathfrak{P}|\mathfrak{p}$. Then $\mathfrak{P}$ is *ramified* over $\mathfrak{p}$ if the exponent $e(\mathfrak{P}/\mathfrak{p}) > 1$. Further, $\mathfrak{p}$ is *ramified* in K if a $\mathfrak{P}$ in K dividing $\mathfrak{p}$ is ramified, otherwise, $\mathfrak{p}$ is *unramified*. If $\mathfrak{P}$ is ramified over $\mathfrak{p}$, the integer $e(\mathfrak{P}/\mathfrak{p}) \geq 2$ is called the *degree of ramification* of $\mathfrak{P}$ over $\mathfrak{p}$.

If $\mathfrak{P}|\mathfrak{p}$, there is an obvious inclusion,

$$\mathfrak{o}/\mathfrak{p} \hookrightarrow \mathcal{O}/\mathfrak{P}$$

of finite fields, taking the coset $x + \mathfrak{p}$ to $x + \mathfrak{P}$. The degree of the field extension

$$f = f(\mathfrak{P}/\mathfrak{p}) = [\mathcal{O}/\mathfrak{P} : \mathfrak{o}/\mathfrak{p}]$$

is called the *residue class degree* of $\mathfrak{P}$ over $\mathfrak{p}$.

EXERCISES

1. Show that $N_{K/k}(\mathfrak{P}) = \mathfrak{p}^{f(\mathfrak{P}/\mathfrak{p})}$.

2. Let $k \subseteq K \subseteq L$ be number fields, $\mathfrak{p}$ a prime in $\mathfrak{o}$, $\mathfrak{P}$ a prime in K dividing $\mathfrak{p}$ and $\mathfrak{Q}$ a prime in L dividing $\mathfrak{P}$. Show that $\mathfrak{Q}|\mathfrak{p}$ and

   (a) $e(\mathfrak{Q}/\mathfrak{p}) = e(\mathfrak{Q}/\mathfrak{P})e(\mathfrak{P}/\mathfrak{p})$ and

   (b) $f(\mathfrak{Q}/\mathfrak{p}) = f(\mathfrak{Q}/\mathfrak{P})f(\mathfrak{P}/\mathfrak{p})$

Theorem 4.6. *Let K/k be an extension of number fields of degree n and $\mathfrak{p}$ a prime in k. Let $\mathfrak{p}\mathcal{O}$ factor in K as in (4.1) and $f_j = f(\mathfrak{P}_j/\mathfrak{p})$. Then*

$$\sum_{j=1}^{g} e_j f_j = n. \tag{4.2}$$

Proof. It is easy to see that $\mathfrak{p} \cap \mathbb{Z}$ is a prime ideal of $\mathbb{Z}$. Let $\mathfrak{p} \cap \mathbb{Z} = p\mathbb{Z} = (p)$. Then $\mathfrak{o}/\mathfrak{p}$ is a finite field of $q = p^d$ elements for some $d \geq 1$. Hence, $N_{k/\mathbb{Q}}(\mathfrak{p}) = (q)$. Taking norm of both sides of (4.1), we have

$$N_{K/\mathbb{Q}}(\mathfrak{p}\mathcal{O}) = N_{K/\mathbb{Q}}(\mathfrak{P}_1^{e_1} \cdots \mathfrak{P}_g^{e_g}). \tag{4.3}$$

But

$$N_{K/\mathbb{Q}}(\mathfrak{p}\mathcal{O}) = N_{k/\mathbb{Q}}(N_{K/k}(\mathfrak{p}\mathcal{O})) = N_{k/\mathbb{Q}}(\mathfrak{p}^n) = (N_{k/\mathbb{Q}}(\mathfrak{p}))^n = (q^n). \tag{4.4}$$

On the other hand,

$$N_{K/\mathbb{Q}}(\mathfrak{P}_1^{e_1} \cdots \mathfrak{P}_g^{e_g}) = N_{K/\mathbb{Q}}(\mathfrak{P}_1^{e_1}) \cdots N_{K/\mathbb{Q}}(\mathfrak{P}_g^{e_g}).$$

Since, $N_{K/\mathbb{Q}}(\mathfrak{P}_j) = (q^{f_i})$,

$$N_{K/\mathbb{Q}}(\mathfrak{P}_1^{e_1} \cdots \mathfrak{P}_g^{e_g}) = (q^{e_1 f_1 + \cdots + e_g f_g}). \tag{4.5}$$

Comparing (4.4) and (4.5), we get

$$q^n = q^{e_1 f_1 + \cdots + e_g f_g}.$$

This shows that

$$n = e_1 f_1 + \cdots + e_g f_g.$$

$\square$

Corollary 4.7. *A prime $\mathfrak{p}$ in k cannot have more than $n = [K : k]$ prime divisors in k.*

Proof. The inequalities $e_j \geq 1$, $f_j \geq 1$ imply $g \leq n$. $\square$

Example 4.8. Take $k = \mathbb{Q}$, $K = \mathbb{Q}(\sqrt{d})$, $d \neq 0, 1$ a square-free integer. There are only three possibilities for a prime in $\mathbb{Q}$ to factor in K.

1. $(p) = \mathfrak{p}^2$ *(ramified)*
2. $(p) = \mathfrak{p}_1 \mathfrak{p}_2$ with $\mathfrak{p}_1 \neq \mathfrak{p}_2$ (splits)
3. $(p) = \mathfrak{p}$ (inert).

In general, there are quite a few possibilities.

Definition 4.9. Let $\mathfrak{p}$ be a prime of k, $[K : k] = n$ and

$$\mathfrak{p}\mathcal{O} = \mathfrak{P}_1^{e_1} \cdots \mathfrak{P}_g^{e_g} \quad (e_j \geq 1)$$

with $\mathfrak{P}_1, \ldots, \mathfrak{P}_g$ distinct. Then

1. if $e = n$ (which is so if and only if $g = 1$, $e = e_1 = n$) we say that $\mathfrak{p}$ is *totally ramified* in K;

2. if $g = n$ (which is equivalent to $e_j = f_j = 1$ for each j) we say that $\mathfrak{p}$ *splits completely* in K and

3. if $f = n$ ($\Leftrightarrow g = 1, e = e_1 = 1$) then $\mathfrak{p}$ is *inert*, or *stays prime* in K.

Remark 4.10. If $n > 2$, there are many other ways for $\mathfrak{p}$ to factor in K. The way in which a given prime $\mathfrak{p}$ of k factors in K is a fundamental problem in algebraic number theory.

4.1   Criterion for Ramification

We now start preparing to show that if K/k is an extension of number fields, the number of primes which ramify in K is finite. In fact, we shall point out exactly which primes in k ramify in K.

Definition 4.11. The *complementary set* of $\mathcal{O}$ relative to $\mathfrak{o}$ is the set

$$\mathcal{O}' = \{\alpha \in K \,|\, \mathrm{Tr}_{K/k}(\alpha\mathcal{O}) \subseteq \mathfrak{o}\}.$$

Theorem 4.12. *The complementary set $\mathcal{O}'$ is a fractional ideal and contains $\mathcal{O}$.*

Proof. It is obvious from the properties of the trace map and definition of $\mathcal{O}'$ that $\mathcal{O}'$ is an $\mathcal{O}$-module and that $\mathcal{O} \subseteq \mathcal{O}'$. All we have to do is to produce a nonzero element d of $\mathfrak{o}$ such that $d\mathcal{O}' \subseteq \mathcal{O}$.

Fix a basis $\alpha_1, \ldots, \alpha_n$ of K over k, consisting of elements of $\mathcal{O}$. If $\alpha \in K$, write

$$\alpha = a_1\alpha_1 + \cdots + a_n\alpha_n \quad (a_j \in k).$$

Then for each $i = 1, \ldots, n$,

$$\mathrm{Tr}_{K/k}(\alpha\alpha_i) = \sum_{j=1}^{n} a_j \, \mathrm{Tr}_{K/k}(\alpha_i\alpha_j) = b_i \in \mathfrak{o},$$

or in the matrix notation

$$(\mathrm{Tr}_{K/k}(\alpha_i\alpha_j)) \begin{pmatrix} a_1 \\ \vdots \\ a_n \end{pmatrix} = \begin{pmatrix} b_1 \\ \vdots \\ b_n \end{pmatrix}.$$

Since the matrix $(\mathrm{Tr}_{K/k}(\alpha_i\alpha_j)) \in GL(n, \mathfrak{o})$, solving by Cramer's rule, we see that $da_j \in \mathfrak{o}$, where $d = \det(\mathrm{Tr}_{K/k}(\alpha_i\alpha_j))$ is a nonzero element of $\mathfrak{o}$. $\square$

Definition 4.13. The integral ideal

$$\mathfrak{D}_{K/k} = \mathcal{O}'^{-1}$$

is called the *different* of K/k.

Definition 4.14. The ideal $\mathfrak{d}_{K/k} = N_{K/k}(\mathfrak{D}_{K/k})$ of $\mathfrak{o}$ is called the *discriminant* of K/k.

The following is also clear from the proof of Theorem 4.12.

Theorem 4.15. *The ideal $\mathfrak{d}_{K/\mathbb{Q}}$ is generated by the discriminant d_K of K, that is, $\mathfrak{d}_{K/\mathbb{Q}} = d_K\mathbb{Z}$.*

The rest of the chapter is devoted to prove that a prime $\mathfrak{P}$ of K is ramified if and only if $\mathfrak{P}|\mathfrak{D}_{K/k}$, and a prime $\mathfrak{p}$ of k is ramified if and only if $\mathfrak{p}|d_{K/k}$. In particular, there are only finitely many primes of $\mathbb{Q}$ which ramify in a number field K. These are exactly the primes which appear in the unique factorization of d_K.

4.2 Review of Commutative Algebra

In this section, we recall some basic facts we need from commutative algebra. For details, see [3].

Localization

Let A be a integral domain with 1, that is, a commutative ring with 1 such that for a, b in A, $ab = 0$ implies that either $a = 0$ or $b = 0$. The ring A is contained, in an obvious way, in its field of fractions K. A subset S of A is a *multiplicative set* if

1. $1 \in S$ but $0 \notin S$, and

2. if $s_1, s_2 \in S$ then $s_1 s_2 \in S$.

The subset

$$S^{-1}A = \left\{ \frac{a}{s} \mid a \in A, s \in S \right\}$$

is a subring of K and contains A as a subring. Every ideal of $S^{-1}A$ is an extension of an ideal $\mathfrak{a}$ of A in the sense that it is generated in $S^{-1}A$ by $\mathfrak{a}$, thus it is of the form $\mathfrak{a}(S^{-1}A) = S^{-1}\mathfrak{a} = \{\frac{a}{s} \mid a \in \mathfrak{a}, s \in S\}$.

Theorem 4.16. *The prime ideals of $S^{-1}A$ are in one-to-one correspondence with prime ideals of A, not intersecting S, via*

$$\mathfrak{p} \to \mathfrak{P} = S^{-1}\mathfrak{p} \to \mathfrak{P} \cap A = \mathfrak{p}.$$

If $\mathfrak{p}$ is a prime ideal of A, then $S = A \setminus \mathfrak{p}$ (the set theoretic difference) is a multiplicative set and $S^{-1}A$ is called the *localization* of A at $\mathfrak{p}$. The ring $A_{\mathfrak{p}} = S^{-1}A$ is a local ring. [A local ring is a commutative ring with 1 having a unique maximal ideal.] The unique maximal ideal of $S^{-1}A$ is $S^{-1}\mathfrak{p} = \mathfrak{p}A_{\mathfrak{p}}$, which is again denoted by $\mathfrak{p}$.

Example 4.17.

1. Take $A = \mathbb{Z}$ and for a fixed prime p, $S = \{p^n \mid n = 0, 1, 2, \ldots\}$. Then $S^{-1}A = \mathbb{Z}\left[\frac{1}{p}\right] = \left\{ \frac{m}{p^r} \mid m \in \mathbb{Z}, r = 0, 1, 2, \ldots \right\}$.

2. Again let $A = \mathbb{Z}$ and p a fixed prime. Suppose $\mathfrak{p} = p\mathbb{Z}$. The localization of A at $\mathfrak{p}$ is the subring

$$A_{\mathfrak{p}} = \left\{ \frac{m}{d} \mid m, d \in \mathbb{Z}, d \geq 1 \text{ and the g.c.d. } (d, p) = 1 \right\}$$

of $\mathbb{Q}$.

Theorem 4.18. *Suppose A is an integral domain, and $\mathfrak{m}$ a maximal ideal of A. Then the fields $A/\mathfrak{m}$ and $A_{\mathfrak{m}}/\mathfrak{m}A_{\mathfrak{m}}$ are isomorphic.*

Proof. It is easy to see that the map $A/\mathfrak{m} \to A_{\mathfrak{m}}/\mathfrak{m}A_{\mathfrak{m}}$ taking the coset $a + \mathfrak{m}$ in $A/\mathfrak{m}$ to $a + \mathfrak{m}A_{\mathfrak{m}}$ of $A_{\mathfrak{m}}/\mathfrak{m}A_{\mathfrak{m}}$ is the required isomorphism. $\square$

Now suppose K/k is an extension of number fields, $\mathfrak{o} = \mathcal{O}_k$, $\mathcal{O} = \mathcal{O}_K$, $\mathfrak{p}$ a prime ideal in k, $S = \mathfrak{o} \setminus \mathfrak{p}$, $A = S^{-1}\mathfrak{o}$, $B = S^{-1}\mathcal{O}$ and $\mathfrak{P} = S^{-1}\mathfrak{p}$.

Theorem 4.19.

i) *The integral closure of A in K is B.*

ii) *B is a finitely generated A-module.*

iii) There is an element π in $\mathfrak{p}$ such that every ideal $\mathfrak{a}$ of A is of the form $\mathfrak{a} = \mathfrak{p}^m = (\pi^m)$. In particular A is a principal ideal domain.

Proof. i) and ii) are trivial. For iii), choose a π in $\mathcal{O}_K$ with $v_\mathfrak{p}(\pi) = 1$. $\square$

EXERCISES

1. Show that B is a *semi-local ring* (a ring having only finitely many maximal ideals). What are these finitely many maximal ideals?

2. Show that B is a PID, and hence a UFD and a Dedekind domain.

4.3   Relative Discriminant for Rings

Let A be a subring of a ring B (all subrings A are assumed to contain 1 of B) such that B is a free A-module of rank n. Recall the definition of the trace map $\mathrm{tr}_{B/A} : B \to A$ as the trace of the matrix for multiplication map m_b by b, for b in B. Note that the matrix of the linear map m_b depends on the basis of B over A, but the trace of this matrix does not.

Definition 4.20. Let A and B be as above. For $\alpha_1, \ldots, \alpha_n$ in B, the element

$$\Delta(\alpha_1, \ldots, \alpha_n) = \det(\mathrm{tr}(\alpha_i \alpha_j))$$

of A is called the *discriminant* of the ordered set $(\alpha_1, \ldots, \alpha_n)$ relative to B/A.

Remark 4.21.

1. The discriminants of two bases of B over A are related by the square of a unit of A, and hence are associates.

2. In general, it is not true that $\Delta(\alpha_1, \ldots, \alpha_n) \neq 0$ if and only if $\alpha_1, \ldots, \alpha_n$ are free over A.

Definition 4.22. The *discriminant of B over A* is the principal ideal of A generated by the discriminant of a basis of B over A. It is denoted by $\mathfrak{d}_{B/A}$.

Let K/k be an extension of number fields with $\mathcal{O} = \mathcal{O}_K$ and $\mathfrak{o} = \mathcal{O}_k$. We know that $\mathcal{O}$ is a finitely generated $\mathfrak{o}$-module, but it may not be free (give a counter-example). So, in general, we need to modify this definition slightly.

Definition 4.23. The *discriminant $\mathfrak{d}_{K/k}$ of K/k* is the ideal of $\mathfrak{o}$ generated by the set of discriminants of all bases of K/k consisting of elements of $\mathcal{O}$.

Remark 4.24.

    1. When the class number h_k of k is one, all the definitions coincide, because every finitely generated module over a principal ideal domain is free.

    2. $\mathfrak{d}_{K/k}$ is a nonzero integral ideal of $\mathfrak{o}$.

4.4   Direct Product of Rings

Suppose $B_1, \ldots, B_r$ are commutative rings with 1. We define their *direct product* as the Cartesian product $B = B_1 \times \cdots \times B_r$, with addition and multiplication taken component-wise. Each B_j may be regarded as a subring of B via the obvious inclusion map, e.g. $B_1 \ni b_1 \to (b_1, 0, \ldots, 0) \in B$.

If A is a subring of each B_j, then A may be regarded as a subring of the direct product $B = B_1 \times \cdots \times B_r$, via the map $A \ni a \to (a, \ldots, a) \in B$.

Theorem 4.25. *Suppose A with 1 is a subring of each B_j and every B_j is a free A-module of rank n_j. Then the direct product $B = B_1 \times \cdots \times B_r$ is a free module of rank $n_1 + \cdots + n_r$. Moreover*

$$\mathfrak{d}_{B/A} = \mathfrak{d}_{B_1/A} \cdots \mathfrak{d}_{B_r/A}. \tag{4.6}$$

Proof. We only need to prove (4.6). To simplify notation, we prove it for $r = 2$. For $r > 2$, the proof is similar.

Put $n_1 = m$ and $n_2 = n$. Let $\alpha_1, \ldots, \alpha_m$ be a basis of B_1 over A and $\beta_1, \ldots, \beta_n$ be a basis of B_2 over A. As A-modules, if we identify B_1 and B_2 with the submodules $B_1 \times \{0\}$ and $\{0\} \times B_2$ of $B = B_1 \times B_2$, then $\{\alpha_1, \ldots, \alpha_m; \beta_1, \ldots, \beta_n\}$ is a basis of B over A. Moreover, for all i, j, we have $\alpha_i \beta_j = 0$. Hence $\Delta(\alpha_1, \ldots, \alpha_m; \beta_1, \ldots, \beta_n)$ is the determinant of the matrix

$$\left(\begin{array}{c|c} \mathrm{tr}_{B_1/A}(\alpha_i \alpha_j) & \\ \hline & \mathrm{tr}_{B_2/A}(\beta_i \beta_j) \end{array} \right)$$

This shows that

$$\Delta(\alpha_1, \ldots, \alpha_m; \beta_1, \ldots, \beta_n) = \Delta(\alpha_1, \ldots, \alpha_m)\Delta(\beta_1, \ldots, \beta_n).$$

Therefore, $\mathfrak{d}_{B/A} = \mathfrak{d}_{B_1/A}\mathfrak{d}_{B_2/A}$. $\qquad\square$

Suppose A is a subring of B. Let $\mathfrak{a}$ be an ideal of A and $\mathfrak{b} = \mathfrak{a}B$ be the ideal of B generated by $\mathfrak{a}$. For α in A and β in B, let $\overline{\alpha}$ and $\overline{\beta}$ denote the residue class of α in $A/\mathfrak{a}$ and that of β in $B/\mathfrak{b}$, respectively.

Theorem 4.26. *Suppose $\{\beta_1, \ldots, \beta_n\}$ is a basis of B over A, such that $\{\overline{\beta}_1, \ldots, \overline{\beta}_n\}$ is a basis of $B/\mathfrak{b}$ over $A/\mathfrak{a}$. Then*

$$\overline{\Delta(\beta_1, \ldots, \beta_n)} = \Delta(\overline{\beta}_1, \ldots, \overline{\beta}_n).$$

Proof. Suppose $\beta \in B$. If the matrix of the map "multiplication by β" with respect to the basis $\beta_1, \ldots, \beta_n$ is (a_{ij}) in $M(n, A)$, then the matrix of the multiplication by $\overline{\beta}$ map with respect to the basis $\overline{\beta}_1, \ldots, \overline{\beta}_n$ is $(\overline{a}_{ij})$ in $M(m, a/\mathfrak{a})$, which shows that

$$\overline{\mathrm{tr}_{B/A}(\beta)} = \mathrm{tr}_{(B/\mathfrak{b})/(A/\mathfrak{a})}(\overline{\beta}).$$

Hence,

$$\overline{\Delta(\beta_1, \ldots, \beta_n)} = \overline{\det(\mathrm{tr}(\beta_i\beta_j))} = \det(\mathrm{tr}(\overline{\beta}_i\overline{\beta}_j)) = \Delta(\overline{\beta}_1, \ldots, \overline{\beta}_n). \qquad \square$$

Theorem 4.27 (Chinese Remainder Theorem). *Suppose A is a commutative ring with 1 and $\mathfrak{a}_1, \ldots, \mathfrak{a}_r$ are pairwise coprime ideals of A, i.e. $\mathfrak{a}_i + \mathfrak{a}_j = A$ for $i \neq j$. Given $a_1, \ldots, a_r$ in A, there is an element x in A, such that*

$$x \equiv a_j \pmod{\mathfrak{a}_j}$$

for all $j = 1, \ldots, r$. Moreover, if for another y in A,

$$y \equiv a_j \pmod{\mathfrak{a}_j}$$

for all $j = 1, \ldots, r$, then

$$x \equiv y \pmod{\mathfrak{a}_1 \cdots \mathfrak{a}_r}.$$

Proof. Put $\mathfrak{b}_i = \prod_{j \neq i} \mathfrak{a}_j$. Then $\mathfrak{a}_j + \mathfrak{b}_j = (1)$. To prove the theorem, now choose b_j in $\mathfrak{b}_j$ such that $b_j \equiv 1 \pmod{\mathfrak{a}_j}$, which implies that $b_j a_j \equiv a_j \pmod{\mathfrak{a}_j}$. Clearly, $b_i a_i \equiv 0 \pmod{\mathfrak{a}_j}$ if $i \neq j$. Hence if we put

$$x = \sum_{i=1}^{n} b_i a_i,$$

then

$$x \equiv a_j \pmod{\mathfrak{a}_j}$$

for all $j = 1, \ldots, n$. If also

$$y \equiv a_j \pmod{\mathfrak{a}_j}$$

for all j, then $x - y$ is in every $\mathfrak{a}_j$, hence in $\mathfrak{a}_1 \cdots \mathfrak{a}_r$. $\qquad \square$

Corollary 4.28. *If $\mathfrak{a}_1, \ldots, \mathfrak{a}_r$ are pairwise coprime ideals of a commutative ring A with 1, and $\mathfrak{a} = \mathfrak{a}_1 \cdots \mathfrak{a}_r$, then the quotient ring $A/\mathfrak{a}$ is isomorphic to the direct product $A/\mathfrak{a}_1 \times \ldots \times A/\mathfrak{a}_r$.*

Proof. By the Chinese Remainder Theorem, the map

$$A/\mathfrak{a} \ni x \bmod \mathfrak{a} \to (x \bmod \mathfrak{a}_1, \dots, x \bmod \mathfrak{a}_r) \in A/\mathfrak{a}_1 \times \cdots \times A/\mathfrak{a}_r$$

is a bijective ring homomorphism. □

Corollary 4.29. *If $\phi(m)$ is the Euler ϕ-function,*

$$\phi(m) = \mathrm{Card}\{a| \, 1 \le a \le m, \, \mathrm{g.c.d.}(a, m) = 1\},$$

then

$$\phi(m) = m \prod_{p|m} \left(1 - \frac{1}{p}\right).$$

Proof. By definition, $\phi(m)$ is the cardinality $\mathrm{Card}\,(\mathbb{Z}/m\mathbb{Z})^\times$ of the group of units of the quotient ring $\mathbb{Z}/m\mathbb{Z}$. If $m = p^d$, then for $1 \le a \le m$, $(a, p) > 1$ if and only if $a = p, 2p, 3p, \dots, p^{d-1}p$.

Hence $\phi(m) = p^d - p^{d-1} = p^d \left(1 - \frac{1}{p}\right)$. If $m = p_1^{d_1} \cdots p_r^{d_r}$, then by Corollary 4.28,

$$(\mathbb{Z}/m\mathbb{Z})^\times \cong (\mathbb{Z}/p_1^{d_1}\mathbb{Z} \times \dots \times \mathbb{Z}/p_r^{d_r}\mathbb{Z})^\times$$
$$\cong (\mathbb{Z}/p_1^{d_1}\mathbb{Z})^\times \times \dots \times (\mathbb{Z}/p_r^{d_r}\mathbb{Z})^\times.$$

Therefore,

$$\phi(m) = \mathrm{Card}\,(\mathbb{Z}/m\mathbb{Z})^\times$$
$$= \mathrm{Card}\,(\mathbb{Z}/p_1^{d_1}\mathbb{Z})^\times \dots \mathrm{Card}\,(\mathbb{Z}/p_r^{d_r}\mathbb{Z})^\times$$
$$= p_1^{d_1}\left(1 - \frac{1}{p_1}\right) \cdots p_r^{d_r}\left(1 - \frac{1}{p_r}\right)$$
$$= m \prod_{p|m}\left(1 - \frac{1}{p}\right).$$

□

4.5 Nilradical

Definition 4.30. An element of a commutative ring A with 1 is *nilpotent* if $a^m = 0$ for some m in $\mathbb{Z}$.

Theorem 4.31. *The set* $\mathrm{nil}(A)$ *of all nilpotent elements of A is an ideal of A.*

The ideal $\mathrm{nil}(A)$ is called the *nilradical* of A.

Proof. Let $x, y \in \mathrm{nil}(A)$. Then for some m, n in $\mathbb{N}$, $x^m = y^n = 0$. If $l = m+n$, then it follows from the Binomial Theorem, that $(x + y)^l = 0$. On the other hand, if $a \in A$, then $(ax)^m = a^m x^m = 0$. This proves that $\mathrm{nil}(A)$ is an ideal of A. $\square$

Theorem 4.32. *The nilradical,* $\mathrm{nil}(A)$, *is the intersection of all prime ideals of A.*

Proof. If x in A is nilpotent, then for some m in $\mathbb{N}$, $x^m = 0$. Hence $x \in \mathfrak{p}$, for all prime ideals $\mathfrak{p}$ of A.

Conversely, suppose x is not nilpotent, that is $x^m \neq 0$ for all m in $\mathbb{N}$. We show that there is at least one prime ideal $\mathfrak{p}$ such that $x \notin \mathfrak{p}$. Let S be the set of ideals $\mathfrak{a}$ of A, such that $x^m \notin \mathfrak{a}$ for all m in $\mathbb{N}$. Clearly, S is not empty, since the zero ideal $(0) \in S$. By Zorn's Lemma, let $\mathfrak{p}$ be a maximal element of S. We shall show that $\mathfrak{p}$ is prime. If not, then there are x, y in $A \backslash \mathfrak{p}$ with xy in $\mathfrak{p}$. Then the ideals $\mathfrak{a} = (\mathfrak{p}, x)$ and $\mathfrak{b} = (\mathfrak{p}, y)$ both properly contain $\mathfrak{p}$. By the choice of $\mathfrak{p}$, for some m, n in $\mathbb{N}$, $x^m \in \mathfrak{a}$, $x^n \in \mathfrak{b}$. This shows that $x^{m+n} \in \mathfrak{a}\mathfrak{b} \subseteq \mathfrak{p}$, implying $\mathfrak{p} \notin S$. This contradiction proves that $\mathfrak{p}$ is prime. $\square$

4.6   Reduced Rings

Definition 4.33. A commutative ring A with 1 is *reduced* if $\mathrm{nil}(A) = (0)$.

Example 4.34.

1. An integral domain is reduced.

2. The product $A_1 \times \ldots \times A_r$ is reduced if all A_j are reduced.

Theorem 4.35. *Suppose K is a number field and $\mathfrak{P}$ a prime ideal of $\mathcal{O} = \mathcal{O}_K$. The quotient ring $\mathcal{O}/\mathfrak{P}^e$ is reduced if and only if $e = 1$.*

Proof. If $e = 1$, then $\mathcal{O}/\mathfrak{P}$ is a field, hence reduced. On the other hand, if $e > 0$, choose π in $\mathfrak{P} - \mathfrak{P}^2$. Then $\pi \neq 0$ in $\mathcal{O}/\mathfrak{P}^e$, but $\pi^e = 0$ in $\mathcal{O}/\mathfrak{P}^e$. Therefore, $\mathcal{O}/\mathfrak{P}^e$ is not reduced. $\square$

Now let A be a subring of a commutative ring B, both with 1. Suppose B is a free A-module of rank n.

Theorem 4.36. *If A is a finite field, then B is reduced if and only if $\mathfrak{d}_{B/A} \neq (0)$.*

Proof. First suppose that B is not reduced, that is, it has a nilpotent element $\alpha \neq 0$. A being a field, α can be completed into a basis $\alpha_1 = \alpha, \alpha_2, \ldots, \alpha_n$ of the vector space B over A. Now all the elements $\alpha_1 \alpha_j$, $j = 1, \ldots, n$ are also nilpotent and since the matrix for a nilpotent element is also nilpotent, its trace is zero (why?). Hence the first row of the matrix $(\text{tr}(\alpha_1 \alpha_j))$ consists of zeros only, which shows that

$$\Delta(\alpha_1, \ldots, \alpha_n) = \det(\text{tr}(\alpha_i \alpha_j)) = 0.$$

Therefore, $\mathfrak{d}_{B/A} = (0)$.

Conversely, suppose B is reduced, i.e. $\text{nil}(B) = \{0\}$. Since $\text{nil}(B)$ is the intersection of all prime ideals and B is finite,

$$(0) = \mathfrak{P}_1 \cap \ldots \cap \mathfrak{P}_r \; (\mathfrak{P}_i \neq \mathfrak{P}_j \text{ for } i \neq j).$$

Every $B/\mathfrak{P}_j$, being a finite integral domain, is a field, hence all $\mathfrak{P}_j$ are maximal and therefore coprime in pairs, and

$$\mathfrak{P}_1 \cap \ldots \cap \mathfrak{P}_r = \mathfrak{P}_1 \ldots \mathfrak{P}_r.$$

By the Corollary 4.28,

$$B = B/(0) = B/\mathfrak{P}_1 \cdots \mathfrak{P}_r \cong B/\mathfrak{P}_1 \times \ldots \times B/\mathfrak{P}_r.$$

By Theorem 4.25,
$$\mathfrak{d}_{B/A} = \mathfrak{d}_{(B/\mathfrak{P}_1)/A} \cdots \mathfrak{d}_{(B/\mathfrak{P}_r)/A}.$$
Since A is a field, each $\mathfrak{d}_{(B/\mathfrak{P}_j)/A} \neq (0)$. Hence $\mathfrak{d}_{B/A} \neq (0)$. $\qquad\square$

4.7 Discriminant and Ramification

Finally we arrive at the main result of this chapter, a criterion for ramification, proved by Dedekind in 1882.

Theorem 4.37. *Suppose K/k is an extension of degree n of number fields. A prime $\mathfrak{p}$ of k ramifies in K if and only if $\mathfrak{p} | \mathfrak{d}_{K/k}$.*

Proof. Let
$$\mathfrak{p}\mathcal{O} = \mathfrak{P}_1^{e_1} \ldots \mathfrak{P}_g^{e_g}$$
be the factorization of $\mathfrak{p}$ into powers of distinct primes in $\mathcal{O}$. Since the ring

$$\mathcal{O}/\mathfrak{p}\mathcal{O} = \mathcal{O}/\mathfrak{P}_1^{e_1} \ldots \mathfrak{P}_g^{e_g} \cong \mathcal{O}/\mathfrak{P}_1^{e_1} \times \ldots \times \mathcal{O}/\mathfrak{P}_g^{e_g},$$

$\mathfrak{p}$ is ramified $\Leftrightarrow$ some $e_j > 1 \Leftrightarrow \mathcal{O}/\mathfrak{P}_j^{e_j}$ is not reduced $\Leftrightarrow \mathcal{O}/\mathfrak{p}\mathcal{O}$ is not reduced $\Leftrightarrow \mathfrak{d}_{(\mathcal{O}/\mathfrak{p}\mathcal{O})/(\mathfrak{o}/\mathfrak{p})} = (0)$. Thus we need to show that $\mathfrak{d}_{(\mathcal{O}/\mathfrak{p}\mathcal{O})/(\mathfrak{o}/\mathfrak{p})} = (0) \Leftrightarrow \mathfrak{p}|\mathfrak{d}_{K/k}$.

Let $S = \mathfrak{o} \setminus \mathfrak{p}$, A the localization of $\mathfrak{o}$ at $\mathfrak{p}$, $B = S^{-1}\mathcal{O}$, $\mathfrak{P} = S^{-1}\mathfrak{p}$, the maximal ideal of A. Since $\mathcal{O}$ is a finitely generated (but not necessarily free) $\mathfrak{o}$-module, B is a finitely generated A-module, generated by the same elements. Now since A is a principal ideal domain, B has a basis over A, easily seen to consist of $n = [K : k]$ elements $\alpha_1, \ldots, \alpha_n$. Since S does not intersect any of the prime ideals of $\mathcal{O}$ lying above $\mathfrak{p}$, we have the following diagram:

$$
\begin{array}{ccc}
\mathcal{O}/\mathfrak{p}\mathcal{O} & \cong & B/\mathfrak{P}B \\
| & & | \\
\mathfrak{o}/\mathfrak{p} & \cong & A/\mathfrak{P}
\end{array}
$$

For β in B, we denote by $\overline{\beta}$ its residue class in $B/\mathfrak{P}B$. The dimension of $\mathcal{O}/\mathfrak{p}\mathcal{O}$ over $\mathfrak{o}/\mathfrak{p}$ is n and so is the dimension of $B/\mathfrak{P}B$ over $A/\mathfrak{P}$. Since $\overline{\alpha}_1, \ldots, \overline{\alpha}_n$ generate $B/\mathfrak{P}B$ over A/B, by comparing dimensions, they must form a basis of $B/\mathfrak{P}B$ over $A/\mathfrak{P}$. Thus by Theorem 4.26 and the diagram above, $\mathfrak{d}_{(\mathcal{O}/\mathfrak{p}\mathcal{O})/(\mathfrak{o}/\mathfrak{p})} = (0)$ if and only if $\Delta(\alpha_1, \ldots, \alpha_n) = 0$. Thus we show that $\mathfrak{p}|\mathfrak{d}_{K/k}$ if and only if $\Delta(\alpha_1, \ldots, \alpha_n) \in \mathfrak{P}$.

First, let $\Delta(\alpha_1, \ldots, \alpha_n) \in \mathfrak{P}$. If $\{\beta_1, \ldots, \beta_n\}$ is a basis of K over k consisting of elements in $\mathcal{O}$, then

$$
\beta_i = \sum_{j=1}^{n} a_{ij}\alpha_j \quad (a_{ij} \in A)
$$

which shows that $\Delta(\beta_1, \ldots, \beta_n) = \det(\mathrm{Tr}(\alpha_i\alpha_j)) \cdot (\det(a_{ij}))^2 \in \mathcal{O} \cap \mathfrak{P} = \mathfrak{p}$. Hence, $\mathfrak{d}_{K/k} \subseteq \mathfrak{p}$, i.e. $\mathfrak{p}|\mathfrak{d}_{K/k}$. Conversely, suppose $\mathfrak{p}|\mathfrak{d}_{K/k}$. If $\alpha_1, \ldots, \alpha_n$ is a basis of B over A, write each $\alpha_j = \beta_j/s$ with β_j in $\mathcal{O}$ and s in S. Then

$$
\Delta(\alpha_1, \ldots, \alpha_n) = \det(\mathrm{tr}(\alpha_i\alpha_j)) = \frac{1}{s^{2n}}\det(\mathrm{tr}(\beta_i\beta_j))
$$

is in $A\mathfrak{d}_{K/k} \subseteq A\mathfrak{p} \subseteq \mathfrak{P}$. $\qquad\qquad\square$

5

Geometry of Numbers

For a commutative ring A with 1, we denote by $A^\times$ its *group of units*, that is $A^\times = \{u \in A \mid vu = 1 \text{ for some } v \text{ in } A\}$. In this chapter, we shall show that the group $\mathcal{O}_K^\times$ of units of a number field K is finitely generated. To motivate, let us take a square-free integer $m > 1$. For the sake of simplicity, let $m \equiv 2, 3$ (mod 4), because then for $K = \mathbb{Q}(\sqrt{m})$, $\mathcal{O}_K = \mathbb{Z}[\sqrt{m}] = \mathbb{Z} \oplus \mathbb{Z}\sqrt{m}$. Now $u = x + y\sqrt{m} \in \mathcal{O}_K^\times$ if and only if the norm

$$N(u) = x^2 - my^2 = \pm 1. \tag{5.1}$$

Thus in the simplest case of the quadratic field $K = \mathbb{Q}(\sqrt{m})$, the determination of $\mathcal{O}_K^\times$ is equivalent to solving the *Pell equation* (5.1).

5.1   Lattices in $\mathbb{R}^n$

If $a \in \mathbb{R}^n$ and $r > 0$, we call the subset

$$B_r(a) = \{x \in \mathbb{R}^n \mid \text{dist}(x, a) = \|x - a\| < r\} \subseteq \mathbb{R}^n$$

the *open ball* of *radius* r, *centered* at a. A subset $X \subseteq \mathbb{R}^n$ is *discrete* if for each a in X, there is an $r > 0$, such that $X \cap B_r(a) = \{a\}$. Consider $\mathbb{R}^n$ as an Abelian group under addition. A *lattice* in $\mathbb{R}^n$ is a discrete subgroup $L \neq \{0\}$ of $\mathbb{R}^n$. Let d be the dimension of the subspace of $\mathbb{R}^n$ spanned by elements of a lattice $L \subseteq \mathbb{R}^n$. Clearly, $d \leq n$. We call d the *rank of the lattice* L. A lattice $L \subseteq \mathbb{R}^n$ is a *full lattice* if its rank is n.

Remark 5.1. Topologically speaking, $L \subseteq \mathbb{R}^n$ is a full lattice if and only if the quotient space $\mathbb{R}^n/L$ is compact.

EXERCISE

Show that $L \subseteq \mathbb{R}^n$ is a lattice if and only if it is a $\mathbb{Z}$-module

$$L = \mathbb{Z}v_1 \oplus \ldots \oplus \mathbb{Z}v_d \tag{5.2}$$

for some vectors $v_1, \ldots, v_d$ in L. The expression (5.2) means that each v in L has a unique representation

$$v = a_1 v_1 + \cdots + a_d v_d$$

with $a_j \in \mathbb{Z}$.

Hint: If $d = 1$, choose $v_1 \neq \mathbf{0}$, a vector in L nearest to $\mathbf{0}$. This is possible, because L is discrete. Then, clearly every vector v in L has a unique representation $v = a v_1$ for a in $\mathbb{Z}$, for if $v = (a + r)v_1$ with $0 < r < 1$, then $r v_1 \in L$, contradicting the choice of v_1. For $d > 1$, use induction on d.

We now give a characterization for a lattice to be full, which is more suitable for our purpose.

Theorem 5.2. *A lattice $L \subseteq \mathbb{R}^n$ is full if and only if there is a bounded set $Y \subseteq \mathbb{R}^n$ such that*

$$\mathbb{R}^n = \cup_{v \in L}(v + Y). \tag{5.3}$$

Here, $v + Y = \{v + y \,|\, y \in Y\}$. Before proving the proposition, we define a useful term.

Definition 5.3. Let

$$L = \mathbb{Z} v_1 \oplus \ldots \oplus \mathbb{Z} v_n \tag{5.4}$$

be a full lattice in $\mathbb{R}^n$. The set

$$P = \{c_1 v_1 + \cdots + c_n v_n \,|\, 0 \leq c_j < 1\} \tag{5.5}$$

is called a *fundamental parallelepiped* of L. It depends on the $\mathbb{Z}$-basis $\{v_1, \ldots, v_n\}$ of L. Clearly P is bounded and

$$\mathbb{R}^n = \cup_{v \in L}(v + P), \tag{5.6}$$

a disjoint union of translates $v + P$ of P by elements of L.

Proof. If L is full, we can take Y to be a fundamental parallelepiped of L.

Conversely, suppose a bounded set $Y \subseteq \mathbb{R}^n$ exists with the property (5.3) and L is not full. We show that this leads to a contradiction.

Let W be the subspace of $\mathbb{R}^n$ spanned by the vectors in L. Then $d = \dim W < n$. Consider $\mathbb{R}^n$ as an inner product space with the dot product of vectors. Choose a unit vector v_{d+1} (by the Gram-Schmidt Process) which is perpendicular to every vector of W. Let $r > 0$ such that $Y \subseteq B_r(\mathbf{0})$. It is easy to see that if $w = a v_{d+1}$ is a vector in $\mathbb{R}^n$ with $a > r$, then $w \notin \cup_{v \in L}(v + Y)$. This is a contradiction. $\square$

5.2   Minkowski's Lemma on Convex Bodies

The Dirichlet's unit theorem asserts that, up to the roots of unity in K, the group $\mathcal{O}_K^\times$ of units of K is a free Abelian group of rank $r = r_1 + r_2 - 1$. [We shall define the non-negative integers r_1 and r_2 in the next section.] It is not very difficult to show that $r \leq r_1 + r_2 - 1$. The harder part that $r = r_1 + r_2 - 1$ follows from the famous lemma of Minkowski on convex bodies.

A subset $X \subseteq \mathbb{R}^n$ is *convex* if for all $\boldsymbol{u}, \boldsymbol{v}$ in X and all real t in the interval $[0, 1]$, the vector $t\boldsymbol{u} + (1 - t)\boldsymbol{v}$ is in X. That is, the line segment joining $\boldsymbol{u}$ to $\boldsymbol{v}$ is entirely in X. It is easy to see that if X is convex in $\mathbb{R}^m$ and Y is convex in $\mathbb{R}^n$, then $X \times Y$ is convex in $\mathbb{R}^{m+n}$. We call $X \subseteq \mathbb{R}^n$ *centrally symmetric* if $\boldsymbol{v} \in X$ implies $-\boldsymbol{v} \in X$.

Let μ be the *Lebesgue measure* on $\mathbb{R}^n$, that is, the measure on $\mathbb{R}^n$, such that for a cube $X \subseteq \mathbb{R}^n$ given by

$$X = \{\boldsymbol{x} = (x_1, \ldots, x_n) \in \mathbb{R}^n \,|\, a_j \leq x_j \leq b_j\},$$

$$\mu(X) = \mathrm{vol}(X) = \prod_{j=1}^n (b_j - a_j).$$

Let L be a full lattice with a fundamental parallelepiped P, as in (5.4) and (5.5). Of course, P depends on the choice of the $\mathbb{Z}$-basis $\{\boldsymbol{v}_1, \ldots, \boldsymbol{v}_n\}$ of L. However, any two $\mathbb{Z}$-bases of L are related by a *unimodular matrix*, that is a matrix of determinant ± 1 with entries in $\mathbb{Z}$. Since $\mu(P)$ is the absolute value of the determinant, whose rows are $\boldsymbol{v}_1, \ldots, \boldsymbol{v}_n$, it follows that the volume $\mu(P)$ of P is independent of the choice of the basis. Thus, we may denote $\mu(P)$ also by $\mu(L)$.

Theorem 5.4 (Minkowski's Lemma). *Suppose $X \subseteq \mathbb{R}^n$ is a bounded, centrally symmetric convex set and $L \subseteq \mathbb{R}^n$ is a full lattice. If $\mu(X) > 2^n \mu(L)$, then X contains a nonzero vector of L.*

Proof. First we show that if $Y \subseteq \mathbb{R}^n$ is a bounded set, such that $\{\boldsymbol{v} + Y \,|\, \boldsymbol{v} \in L\}$ is a family of disjoint subsets of $\mathbb{R}^n$, then $\mu(Y) \leq \mu(P)$, where P is a fundamental parallelepiped of L. This is almost immediate, because writing Y as the disjoint union

$$Y = \cup_{\boldsymbol{v} \in L} Y \cap (\boldsymbol{v} + P),$$

we have by (5.6), $\mu(Y) = \sum_{\boldsymbol{v} \in L} \mu(Y \cap (\boldsymbol{v} + P))$.

Since μ is translation invariant, $\mu(Y \cap (\boldsymbol{v} + P)) = \mu((-\boldsymbol{v} + Y) \cap P)$. Hence $\mu(Y) = \sum_{\boldsymbol{v} \in L} \mu((-\boldsymbol{v} + Y) \cap P) \leq \mu(P)$, because the sets $-\boldsymbol{v} + Y$ are also pairwise disjoint.

Take now $Y = \frac{1}{2}X = \left\{\frac{1}{2}\boldsymbol{v} \mid \boldsymbol{v} \in X\right\}$. It is given that $\mu(Y) = \frac{1}{2^n} \cdot \mu(X) > \mu(P)$. Hence the translates $\boldsymbol{v} + Y$, $\boldsymbol{v} \in L$ of Y are not pairwise disjoint, that is, there are two distinct vectors $\boldsymbol{v}_1, \boldsymbol{v}_2$ in L, and $\boldsymbol{u}_1, \boldsymbol{u}_2$ in X, such that $\boldsymbol{v}_1 + \frac{1}{2}\boldsymbol{u}_1 = \boldsymbol{v}_2 + \frac{1}{2}\boldsymbol{u}_2$. Since X is centrally symmetric and convex, this shows that $\boldsymbol{v}_1 - \boldsymbol{v}_2 = \frac{1}{2}\boldsymbol{u}_2 - \frac{1}{2}\boldsymbol{u}_1 \in X$. Since $\boldsymbol{v}_1 - \boldsymbol{v}_2$ is a nonzero element of L, we are done. □

Remark 5.5. In Minkowski's Lemma, the hypothesis $\mu(X) > 2^n\mu(L)$ may be replaced by $\mu(X) \geq 2^n\mu(L)$, if X is compact.

Corollary 5.6. *Suppose $Y \subseteq \mathbb{R}^n$ is a measurable set. If $\cup_{\boldsymbol{v} \in L}(\boldsymbol{v} + Y) = \mathbb{R}^n$, then $\mu(Y) \geq \mu(P)$.*

5.3 Logarithmic Embedding

Suppose $K \subseteq \mathbb{C}$ is a number field of degree n over $\mathbb{Q}$. Consider a ring homomorphism $\sigma : K \to \mathbb{C}$. We require that $\sigma(1) = 1$. Hence $\sigma_{|\mathbb{Q}} = 1_{\mathbb{Q}}$, the identity map on $\mathbb{Q}$. Such a σ is clearly injective. [Its kernel $\mathrm{Ker}(\sigma)$ is an ideal of the field K, which can only be $\{0\}$ or K.] Hence, we call σ a $\mathbb{Q}$-*isomorphism* of K into $\mathbb{C}$. There are exactly n $\mathbb{Q}$-isomorphisms of K into $\mathbb{C}$. To see this, write $K = \mathbb{Q}(\alpha)$. If σ is a $\mathbb{Q}$-isomorphism of K into $\mathbb{C}$, it is determined by $\sigma(\alpha)$, which is a conjugate of α. But there are exactly n conjugates of α over $\mathbb{Q}$.

One may regard such a $\sigma : K \to \mathbb{C}$ also an injective linear transformation of vector spaces, when K and $\mathbb{C}$ are viewed as vector spaces over $\mathbb{Q}$. Unless stated to the contrary $\sigma : K \to \mathbb{C}$ will be a $\mathbb{Q}$-isomorphism.

If $\sigma(K) \subseteq \mathbb{R}$, we call σ a *real imbedding*, otherwise it is a *complex imbedding*. If σ is complex, the map $\overline{\sigma} : K \to \mathbb{C}$, given by $\overline{\sigma}(x) = \overline{\sigma(x)}$ is also a $\mathbb{Q}$-isomorphism. Thus, the complex $\mathbb{Q}$-isomorphisms occur in pairs. We shall denote the real $\mathbb{Q}$-isomorphisms of K into $\mathbb{C}$ by $\sigma_1, \ldots \sigma_{r_1}$ and the r_2 pairs of complex ones by $\sigma_{r_1+1}, \overline{\sigma_{r_1+1}}; \ldots; \sigma_{r_1+r_2}, \overline{\sigma_{r_1+r_2}}$. In particular, $n = r_1 + 2r_2$.

Consider $\mathbb{C}$ as a vector space of dimension two over $\mathbb{R}$ with $\{1, i\}$ as the standard basis. If $z = x + iy \in \mathbb{C}$, the multiplication by z is a linear transformation of $\mathbb{C}$ into itself over $\mathbb{R}$. Its matrix relative to the basis $\{1, i\}$ is easily seen to be $T = \begin{pmatrix} x & y \\ -y & x \end{pmatrix}$ with determinant

$$\det(T) = x^2 + y^2 = |z|^2. \tag{5.7}$$

If we identify $\mathbb{C}$, as a vector space over $\mathbb{R}$ with $\mathbb{R}^2$, via the map $x + iy \to \begin{pmatrix} x \\ y \end{pmatrix}$, then $\mathbb{R}^{r_1} \times \mathbb{C}^{r_2} \cong \mathbb{R}^n$.

For a fixed vector $\boldsymbol{\alpha} = (x_1, \ldots, x_{r_1}; z_1, \ldots, z_{r_2})$ of $V = \mathbb{R}^{r_1} \times \mathbb{C}^{r_2}$, the map

$$\boldsymbol{\beta} = (x_1', \ldots, x_{r_1}'; z_1', \ldots, z_{r_2}') \to (x_1 x_1', \ldots, x_{r_1} x_{r_1}'; z_1 z_1', \ldots, z_{r_2} z_{r_2}')$$

defines an $\mathbb{R}$-linear map from V to itself. The determinant of its matrix, which we also denote by $\boldsymbol{\alpha}$, is

$$\det(\boldsymbol{\alpha}) = x_1 \cdots x_{r_1} |z_1|^2 \cdots |z_{r_2}|^2, \tag{5.8}$$

in view of (5.7).

We now define a $\mathbb{Q}$-linear map

$$\rho : K \to V = \mathbb{R}^{r_1} \times \mathbb{C}^{r_2} \cong \mathbb{R}^n$$

by

$$\rho(\alpha) = (\sigma_1(\alpha), \ldots, \sigma_{r_1}(\alpha); \sigma_{r_1+1}(\alpha), \ldots, \sigma_{r_1+r_2}(\alpha)).$$

It is clear from (5.8) that

$$\det(\rho(\alpha)) = N(\alpha), \tag{5.9}$$

where N is the norm $N_{K/\mathbb{Q}}$, as defined in Chapter 3.

Theorem 5.7. *If $\alpha_1, \ldots, \alpha_n$ is a basis of K over $\mathbb{Q}$, then $\{\rho(\alpha_1), \ldots, \rho(\alpha_n)\}$ is a basis of $\mathbb{R}^n$ over $\mathbb{R}$. In particular, $\rho(\mathcal{O}_K)$ is a full lattice in $\mathbb{R}^n$.*

Proof. All we need to do is show that the determinant

$$d = \begin{vmatrix} x_1^{(1)} \ldots x_{r_1}^{(1)} & x_{r_1+1}^{(1)} & y_{r_1+1}^{(1)} & \cdots & x_{r_1+r_2}^{(1)} & y_{r_1+r_2}^{(1)} \\ \vdots & & & & & \\ x_1^{(n)} \ldots x_{r_1}^{(n)} & x_{r_1+1}^{(n)} & y_{r_1+1}^{(n)} & \cdots & x_{r_1+r_2}^{(n)} & y_{r_1+r_2}^{(n)} \end{vmatrix} \neq 0,$$

where

$$\sigma_r(\alpha_s) = \begin{cases} x_s^{(r)} & \text{if } 1 \leq r \leq r_1 \\ x_s^{(r)} + iy_s^{(r)} & \text{if } r_1 < r \leq r_1 + r_2. \end{cases}$$

For $r_1 < r \leq r_1 + r_2$, $x_s^{(r)} = \frac{1}{2}(\sigma_r(\alpha_s) + \overline{\sigma}_r(\alpha_s))$ and $y_s^{(r)} = \frac{1}{2i}(\sigma_r(\alpha_s) - \overline{\sigma}_r(\alpha_s))$. Substituting this in the above determinant and performing the obvious column operations, one gets

$$d = \frac{1}{(-2i)^{r_2}} \begin{vmatrix} \sigma_1(\alpha_1) \ldots \sigma_{r_1}(\alpha_1) & \sigma_{r_1+1}(\alpha_1) & \overline{\sigma}_{r_1+1}(\alpha_1) & \cdots \\ \vdots & & & \\ \sigma_1(\alpha_n) \ldots \sigma_{r_1}(\alpha_n) & \sigma_{r_1+1}(\alpha_n) & \overline{\sigma}_{r_1+1}(\alpha_n) & \cdots \end{vmatrix}$$

$$= \frac{1}{(-2i)^{r_2}} \cdot D(\alpha_1, \ldots, \alpha_n).$$

The determinant $D(\alpha_1, \ldots, \alpha_n)$ is related to the discriminant d^* of the basis $\alpha_1, \ldots, \alpha_n$ of K over $\mathbb{Q}$ by $d^* = D^2(\alpha_1, \ldots, \alpha_n)$. Since $d^* \neq 0$, then likewise $d \neq 0$. $\qquad \square$

Corollary 5.8. *The volume of the fundamental parallelepiped of the full lattice $\rho(\mathcal{O}_K^\times)$ is $\frac{1}{2^{r_2}} \sqrt{|d_K|}$, d_K being the discriminant of the number field K.*

Proof. If $\mathcal{O}_K = \mathbb{Z}\alpha_1 \oplus \cdots \oplus \mathbb{Z}\alpha_n$, then $d_K = D^2(\alpha_1, \ldots, \alpha_n)$. If P is the fundamental parallelepiped of $\rho(\mathcal{O}_K^\times)$, then the volume $\mu(P)$ of P is, up to sign, the determinant whose rows are $\rho(\alpha_1), \ldots, \rho(\alpha_n)$. Hence

$$\mu(P) = \frac{1}{2^{r_2}}|D(\alpha_1, \ldots, \alpha_n)| = \frac{1}{2^{r_2}} \sqrt{|d_K|}.$$

We now define a map $\lambda : K^\times \to \mathbb{R}^{r_1+r_2}$, called the *logarithmic imbedding* of $K^\times$ in $\mathbb{R}^{r_1+r_2}$ by

$$\lambda(\alpha) = (\log|\sigma_1(\alpha)|, \ldots, \log|\sigma_{r_1}(\alpha)|; \log|\sigma_{r_1+1}(\alpha)|^2, \ldots, \log|\sigma_{r_1+r_2}(\alpha)|^2).$$

[Here log is the natural logarithm to the base e.] Clearly $\lambda(\alpha\beta) = \lambda(\alpha) + \lambda(\beta)$, hence λ is a group homomorphism from the multiplicative group $K^\times$ to the additive group $\mathbb{R}^{r_1+r_2}$. □

Theorem 5.9. *$\lambda(\mathcal{O}_K^\times)$ is a lattice in $\mathbb{R}^{r_1+r_2}$.*

Proof. We only need to show that the subgroup $\lambda(\mathcal{O}_K^\times)$ of the additive group $\mathbb{R}^{r_1+r_2}$ is discrete. For $t > 0$, consider the subset

$$X = \{(\log|\sigma_1(\alpha)|, \ldots, \log|\sigma_{r_1}(\alpha)|; \log|\sigma_{r_1+1}(\alpha)|^2, \ldots,$$
$$\log|\sigma_{r_1+r_2}(\alpha)|^2)|\, \alpha \in K^\times\} \cap B_t(\mathbf{0})$$

of $\mathbb{R}^{r_1+r_2}$. For a vector in X, we then have

$$\log|\sigma_j(\alpha)| < t \text{ if } 1 \leq j \leq r_1$$

and

$$\log|\sigma_j(\alpha)|^2 < t \text{ if } r_1 < j \leq r_1 + r_2.$$

This means

$$|\sigma_j(\alpha)| < e^t \text{ if } 1 \leq j \leq r_1$$

and

$$|\sigma_j(\alpha)| < e^{t/2} \text{ if } r_1 < j \leq r_1 + r_2.$$

Since, by Theorem 5.7, $\rho(\mathcal{O}_K)$ is a lattice in $\mathbb{R}^n$, there are only finitely many α in $\mathcal{O}_K^\times$ with $\lambda(\alpha)$ in X. □

Theorem 5.10. *The kernel, $\ker(\lambda)$ of the restriction map, also denoted by $\lambda : \mathcal{O}_K^\times \to \mathbb{R}^{r_1+r_2}$ is a finite group, and consists of precisely the roots of unity in K.*

Remark 5.11. We denote the group of roots of unity in K by W_K. The theorem asserts that $\ker(\lambda) = \{\alpha \in \mathcal{O}_K^\times \mid \lambda(\alpha) = \mathbf{0}\}$ is the finite group W_K.

Proof. If $\zeta \in W_K$, then $\sigma(\zeta) \in W_K$ for every $\mathbb{Q}$-isomorphism σ of K into $\mathbb{C}$, hence $|\sigma(\zeta)| = 1$. Therefore, $\lambda(\zeta) = \mathbf{0}$, which shows that $W_K \subseteq \ker(\lambda)$. Conversely, every component of the vector $\rho(\alpha)$ for α in $\ker(\lambda)$ has absolute value 1. Since ρ is injective and $\rho(\mathcal{O}_K^\times)$ is discrete, $\ker(\lambda)$ is finite. Hence for α in $\ker(\lambda)$, the powers $\alpha, \alpha^2, \alpha^3, \ldots$ of α repeat, i.e. for some $t > s$, $\alpha^t = \alpha^s$, which shows that $\alpha^m = 1$ for $m = t - s$. Hence $\ker(\lambda) \subseteq W_K$. $\square$

Theorem 5.12 (Dirichlet 1846). *Let $r = r_1 + r_2 - 1$. The group of units $\mathcal{O}_K^\times$ is isomorphic to $W_K \times \mathbb{Z}^r$. In other words, there are r units $u_1, \ldots, u_r$ in $\mathcal{O}_K^\times$ such that every unit $u \in \mathcal{O}_K^\times$ has a unique representation*

$$u = \zeta u_1^{a_1} \cdots u_r^{a_r}$$

with ζ in W_K and a_j in $\mathbb{Z}$.

Dirichlet's unit theorem, Theorem 5.12, now follows at once from Theorem 5.10 and the first isomorphism theorem in group theory, if we prove the following fact.

Theorem 5.13. *The rank r of the lattice $\lambda(\mathcal{O}_K^\times)$ is given by $r = r_1 + r_2 - 1$.*

Proof. The units u of $\mathcal{O}_K$ are characterized by $N(u) = \prod_\sigma \sigma(u) = \pm 1$, where the product is over all $\mathbb{Q}$-isomorphisms of K into $\mathbb{C}$. Therefore for $\alpha \in \mathcal{O}_K^\times$, $\log|\sigma_1(\alpha)| + \cdots + \log|\sigma_{r_1}(\alpha)| + \log|\sigma_{r_1+1}(\alpha)|^2 + \cdots + \log|\sigma_{r_1+r_2}(\alpha)|^2 = \log|\sigma_1(\alpha) \cdots \sigma_{r_1}(\alpha) \cdot \sigma_{r_1+1}(\alpha) \overline{\sigma}_{r_1+1}(\alpha) \cdots \sigma_{r_1+r_2}(\alpha) \cdot \overline{\sigma}_{r_1+r_2}(\alpha)| = \log|N(\alpha)| = 0$. Hence $\lambda(\mathcal{O}_K^\times)$ is contained in the hyperplane H of $\mathbb{R}^{r_1+r_2}$ defined by the equation

$$\lambda_1 + \cdots + \lambda_{r_1+r_2} = 0.$$

Since $\dim H = r_1 + r_2 - 1$, the rank of $\lambda(\mathcal{O}_K^\times) \leq r_1 + r_2 - 1$.

We use Minkowski's Lemma to show that $r = r_1 + r_2 - 1$, that is, $\lambda(\mathcal{O}_K^\times)$ is a full lattice in H.

For $\boldsymbol{\alpha} = (x_1, \ldots, x_{r_1}; z_1, \ldots, z_{r_2})$ in $V = \mathbb{R}^{r_1} \times \mathbb{C}^{r_2}$, we define its norm $N(\boldsymbol{\alpha})$ by

$$N(\boldsymbol{\alpha}) = x_1 \cdots x_{r_1} |z_1|^2 \cdots |z_{r_2}|^2.$$

Let S be the subset of V of elements $\boldsymbol{\alpha}$ with $|N(\boldsymbol{\alpha})| = 1$. The map $l : S \to \mathbb{R}^{r_1+r_2}$ defined by

$$l(\boldsymbol{\alpha}) = (\log|x_1|, \ldots, \log|x_{r_1}|; \log|z_1|^2, \ldots, \log|z_{r_2}|^2)$$

has the following properties.

1. *The image $l(S)$ of S is the hyperplane $H \subseteq \mathbb{R}^{r_1+r_2}$ given by the equation*

$$\lambda_1 + \cdots + \lambda_{r_1+r_2} = 0.$$

2. *The image $l(X)$ of a bounded set $X \subseteq S$ is also bounded.*

(1) is obvious. To prove (2), let $|x_j|, |z_j|^2 < C$ for all j. Then the coordinates of points in $l(X)$ satisfy the inequalities $\log |x_j|, \log |z_j|^2 < \log C$, and hence are bounded above. On the other hand, for $\alpha = (x_1, \ldots, x_{r_1}; z_1, \ldots, z_{r_2})$ in $X \subseteq S$, $|N(\alpha)| = 1$, hence

$$\sum_{j=1}^{r_1} \log |x_j| + \sum_{j=1}^{r_2} \log |z_j|^2 = 0.$$

This gives

$$\log |x_i| = -\left(\sum_{j \neq i} \log |x_j| + \sum \log |z_j|^2 \right)$$

$$> -(r_1 + r_2 - 1) \log C$$

and

$$\log |z_i|^2 = -\left(\sum \log |x_j| + \sum_{j \neq i} \log |z_j|^2 \right)$$

$$> -(r_1 + r_2 - 1) \log C.$$

Hence the coordinates of a point in $l(X)$ are also bounded below. Hence $l(X)$ is bounded.

To prove that $\lambda(\mathcal{O}_K^\times) = l(\rho(\mathcal{O}_K^\times))$ is a full lattice in the hyperplane H, by Theorem 5.2, it is enough to find a bounded set in H whose translates by the elements of $\lambda(\mathcal{O}_K^\times)$ cover H. Since $\lambda(S) = H$ and for $X \subseteq S$, $l(\rho(\mathcal{O}_K^\times) \cdot X) = l(\rho(\mathcal{O}_K^\times)) + l(X)$, it suffices to find a bounded set $Y \subseteq S$, such that

$$S = Y \cdot \rho(\mathcal{O}_K^\times) = \{ y \cdot \rho(u) \mid y \text{ in } Y, u \in \mathcal{O}_K^\times \}. \tag{5.10}$$

[Here the product $y \cdot \rho(u)$ is componentwise in $V = \mathbb{R}^{r_1} \times \mathbb{C}^{r_2}$.] We take for $a > 0$

$$X = \{(x_1, \ldots, x_{r_1}; z_1, \ldots, z_{r_2}) \in S \mid |x_j| \text{ and } |z_j| < a\}$$

It is easy to see that its volume

$$\mu(X) = \int_{-a}^{a} \cdots \int_{-a}^{a} dx_1 \cdots dx_{r_1} \int_{|z_{r_2}|<a} \cdots \int_{|z_1|<a} dz_1 \cdots dz_{r_2} = (2a)^{r_1}(\pi a^2)^{r_2} = 2^{r_1}\pi^{r_2}a^n.$$

To apply Minkowski's Lemma, we choose a large enough so that $\mu(X)$ is bigger than 2^n times the volume of the fundamental parallelepiped of $\rho(\mathcal{O}_K^\times)$. We already computed (see Corollary 5.8) the volume of the fundamental parallelepiped to be $\frac{1}{2^{r_2}}\sqrt{|d_K|}$. Hence our choice of a is such that

$$\mu(X) > 2^n \cdot \frac{1}{2^{r_2}}\sqrt{|d_K|},$$

i.e.

$$a^n > \left(\frac{2}{\pi}\right)^{r_2}\sqrt{|d_K|}.$$

To find Y for which (5.10) holds, let $\boldsymbol{x} \in S$. We replace the full lattice $L = \rho(\mathcal{O}_K) \subseteq V = \mathbb{R}^{r_1} \times \mathbb{C}^{r_2}$ by $M = \boldsymbol{x} \cdot L$. Since the map, $\boldsymbol{y} \to \boldsymbol{x} \cdot \boldsymbol{y}$ (componentwise multiplication in V) is an invertible linear map, M is also a full lattice in V. Moreover the determinant of this linear map is equal to $N(\boldsymbol{x}) = \pm 1$, which implies that the volume of the fundamental parallelepiped is invariant under this map. Therefore, the volume of the fundamental parallelepiped of $\boldsymbol{x}M$ is also $\frac{1}{2^{r_2}}\sqrt{|d_K|}$. By Minkowski's Lemma, applied to the set X and the lattice $M = \boldsymbol{x}\rho(\mathcal{O}_K)$, we find $\alpha \neq 0$ in $\mathcal{O}_K$ such that $\boldsymbol{y} = \boldsymbol{x} \cdot \rho(\alpha)$ is a nonzero element of X. Since $N(\boldsymbol{x}) = \pm 1$,

$$|N(\alpha)| = |\prod_\sigma \sigma(\alpha)| = |N(\rho(\alpha)| = |N(\boldsymbol{y})| < a^n.$$

[The product $\prod_\sigma \sigma(\alpha)$ is over all the $\mathbb{Q}$-isomorphisms σ of K into $\mathbb{C}$.] In Chapter 3, we proved that there are only finitely many ideals of bounded norm. In particular, there are only finitely many principal ideals of bounded norm. Since the norm $N((\alpha))$ of a principal ideal is the same as the absolute value of the norm $N(\alpha) = \prod_\sigma \sigma(\alpha)$ of α, this implies that, up to multiplication by units of $\mathcal{O}_K$, there are only finitely many α in $\mathcal{O}_K$ with $|N(\alpha)| < a^n$. Call them $\alpha_1, \dots, \alpha_m$. Therefore, $\alpha = u\alpha_j$ for some j $(1 \leq j \leq m)$ and hence

$$\boldsymbol{y} = \boldsymbol{x}\rho(\alpha) = \boldsymbol{x}\rho(u)\rho(\alpha_j), \text{ i.e. } \boldsymbol{x} = \rho(u^{-1})\rho(\alpha_j^{-1})\boldsymbol{y}.$$

Since $\rho(\alpha_j^{-1})\boldsymbol{y}$ and $\boldsymbol{x}$ both have norm ± 1, $\rho(\alpha_j^{-1})\boldsymbol{y} \in S \cap \rho(\alpha_j^{-1})X$, which shows that $\boldsymbol{x} \in \lambda(\mathcal{O}_K^\times) \cdot (S \cap \rho(\alpha_j^{-1})X)$ for some j $(1 \leq j \leq m)$.

Now we put

$$Y = \cup_{j=1}^m (S \cap \rho(\alpha_j^{-1})X).$$

Since a and α_j were independent of $\boldsymbol{x}$, every element $\boldsymbol{x}$ of S is in $\rho(\mathcal{O}_K^\times) \cdot Y$. Since each $\rho(\alpha_j^{-1})X$ is bounded, so is Y. This proves (5.10), concluding the proof of Dirichlet's unit theorem. $\qquad\square$

Remark 5.14. In Dirichlet's unit theorem, the units $u_1, \dots, u_r$, which generate the free part of the group $\mathcal{O}_K^\times$, are called *fundamental units*. The set $\{u_1, \dots, u_r\}$ of fundamental units is not unique. What is unique is the cardinality r of this set. We call r the *rank* of $\mathcal{O}_K^\times$.

5.4   Units of a Quadratic Field

Let $K = \mathbb{Q}(\sqrt{d})$ ($d \neq 0, 1$, a square-free integer) be a quadratic field. We call K a *real quadratic field* or an *imaginary quadratic field* according as $d > 0$ or $d < 0$. If K is an imaginary quadratic field, then $r_1 = 0$, $r_2 = 1$, so $r = r_1 + r_2 - 1 = 0$. In this case, $\mathcal{O}_K^\times = W_K$, the roots of unity in K. We leave it as an exercise to determine this finite group W_K.

For the real quadratic field, $r = 1$ and the group of units is given by the following corollary.

Corollary 5.15. *If $d > 1$ is a square-free integer and $K = \mathbb{Q}(\sqrt{d})$, then the group*

$$\mathcal{O}_K^\times \cong \{\pm 1\} \times \mathbb{Z}.$$

In particular, the Pell equation $x^2 - dy^2 = 1$ has infinitely many solutions in integers.

EXERCISES

1. Determine the structure of $\mathcal{O}_K^\times$ when $[K : \mathbb{Q}] = 3$ and 4.

2. Use Dirichlet's unit theorem to find all integer solutions of $5x^2 - 5y^2 = y^4$.

5.5   Estimates on the Discriminant

In view of Dedekind's theorem on ramification, it is clear that the discriminant is the most important invariant of the ground field k for studying the ramification of primes of k in a finite extension K/k of number fields. Thus a minimal knowledge of this invariant is absolutely essential for studying the arithmetic in such extensions. In this section, we obtain some classical results on the discriminant d_K for $K/\mathbb{Q}$.

For non-negative integers d and m, let $n = d + 2m$. We put $V = \mathbb{R}^d \times \mathbb{C}^m$, which when viewed as a vector space over $\mathbb{R}$ has dimension n. The Lebesgue measure on V will be denoted by μ. For $n = 1$, it is length, for $n = 2$, it is area and for $n = 3$, it is volume, etc. We identify $\mathbb{C}$ with $\mathbb{R}^2$, via the map $\mathbb{C} \ni z = x + iy \rightarrow (x, y) \in \mathbb{R}^2$. For a real number $a > 0$, we put

$$S_a(d, m) = \{(x_1, \ldots, x_d; z_1, \ldots, z_m) \in V \mid \sum_{j=1}^{d} |x_j| + 2 \sum_{j=1}^{m} |z_j| \leq a\}.$$

Clearly, $S_a(d, m)$ is a compact, convex, measurable and centrally symmetric subset of $\mathbb{R}^n$.

Theorem 5.16.
$$\mu(S_a(d, m)) = 2^d \left(\frac{\pi}{2}\right)^m \cdot \frac{a^n}{n!}. \tag{5.11}$$

Proof. We use double induction on d and m. We have
$$\mu(S_a(1, 0)) = \mu([-a, a]) = 2a,$$
which agrees with (5.11). Next,
$$\mu(S_a(0, 1)) = \pi \left(\frac{a}{2}\right)^2,$$
which also agrees with (5.11).

To compute $\mu(S_a(d+1, m))$, we integrate last with respect to the $(d+1)^{th}$ variable x, and use the symmetry. So by the induction hypothesis,

$$\mu(S_a(d+1, m)) = 2 \int_{-a}^{a} \mu(S_a(d, m)) dx$$

$$= 2 \cdot 2^d \left(\frac{\pi}{2}\right)^m \cdot \frac{1}{n!} \int_0^a (a - x)^n dx$$

$$= 2^{d+1} \left(\frac{\pi}{2}\right)^m \frac{a^{n+1}}{(n+1)!}.$$

Because $n = d + 2m$, if (5.11) is true for d, then it is true for $d + 1$ also.

To compute $\mu(S_a(d, m+1))$, note that $S_a(d, m+1)$ is defined by

$$\sum_{j=1}^{d} |x_j| + 2 \sum_{j=1}^{m} |z_j| + 2|z| \leq a.$$

If we fix the $(m+1)^{th}$ variable until the end, by the induction hypothesis,

$$\mu(S_{a-2|z|}(d, m)) = 2^d \left(\frac{\pi}{2}\right)^m \frac{(a - 2|z|)^n}{n!}.$$

Finally, integrating with respect to the last variable z,

$$\mu(S_a(d, m+1)) = 2^d \left(\frac{\pi}{2}\right)^m \frac{1}{n!} \int_{|z| \leq \frac{a}{2}} (a - 2|z|)^n d\mu(z).$$

We put $z = re^{i\theta}$, $0 \le r \le \frac{a}{2}$, $0 \le \theta \le 2\pi$. Then

$$\int_{|z| \le \frac{a}{2}} (a - 2|z|)^n d\mu(z) = \int_0^{\frac{a}{2}} \int_0^{2\pi} (a - 2r)^n r \, dr \, d\theta$$

$$= 2\pi \int_0^{\frac{a}{2}} (a - 2r)^n r \, dr$$

$$= \frac{\pi}{2} \cdot \frac{a^{n+2}}{(n+1)(n+2)}.$$

Hence,

$$\mu(S_a(d, m+1)) = 2^d \left(\frac{\pi}{2}\right)^{m+1} \cdot \frac{a^{d+2(m+1)}}{(d+2(m+1))!}.$$

This completes the proof by double induction. $\qquad\qquad\square$

Now let K be a number field of degree $n = r_1 + 2r_2$. Let $\sigma_1, \ldots, \sigma_{r_1}$ be the r_1 real $\mathbb{Q}$-isomorphisms of K into $\mathbb{R}$, whereas $\sigma_{r_1+1}, \overline{\sigma}_{r_1+1}; \ldots; \sigma_{r_1+r_2}, \overline{\sigma}_{r_1+r_2}$ the r_2 pairs of complex $\mathbb{Q}$-isomorphisms of K into $\mathbb{C}$. In particular, $n = r_1 + 2r_2$.

Again, let

$$\rho : K \to V = \mathbb{R}^{r_1} \times \mathbb{C}^{r_2} \cong \mathbb{R}^n$$

be the canonical imbedding of K in V. It is given by

$$\rho(\alpha) = (\sigma_1(\alpha), \ldots, \sigma_{r_1}(\alpha); \sigma_{r_1+1}(\alpha), \ldots, \sigma_{r_1+r_2}(\alpha)).$$

We shall need the following fact.

Theorem 5.17. *For non-negative real numbers $x_1, \ldots, x_n$, show that*

$$\sqrt[n]{x_1 \cdots x_n} \le \frac{x_1 + \cdots + x_n}{n},$$

i.e. the geometric mean never exceeds the arithmetic mean.

Proof. Use the Lagrange method of multipliers to maximize the function $f(x_1, \ldots, x_n) = (x_1 \ldots x_n)^{1/n}$ subject to the condition $x_1 + \cdots + x_n = c$, a constant. $\qquad\qquad\square$

Theorem 5.18. *Suppose K is a number field with $[K : \mathbb{Q}] = n = r_1 + 2r_2$. If $\mathfrak{a} \ne (0)$ is an integral ideal in K, then $\mathfrak{a}$ contains a nonzero element α, such that*

$$|N(\alpha)| \le \left(\frac{4}{\pi}\right)^{r_2} \frac{n!}{n^n} \sqrt{|d_K|} \cdot N(\mathfrak{a}).$$

Proof. Recall that for a lattice,

$$L = \mathbb{Z}\boldsymbol{v}_1 \oplus \ldots \oplus \mathbb{Z}\boldsymbol{v}_n$$

in $\mathbb{R}^n$, the volume $\mu(P)$ of its fundamental parallelepiped

$$P = \{a_1\boldsymbol{v}_1 + \cdots + a_n\boldsymbol{v}_n | 0 \le a_j < 1\}$$

is independent of the $\mathbb{Z}$-basis $\{\boldsymbol{v}_1, \ldots, \boldsymbol{v}_n\}$ of L. Hence we may denote it by $\mu(L)$ and call it the *volume* of L. If $M \subseteq L$ is a full sublattice, of index $[L : M] = m$, then a fundamental parallelepiped of M is a disjoint union of m translates of a fundamental parallelepiped of L. Since μ is translation invariant,

$$\mu(M) = m\mu(L).$$

In particular, if $\mathfrak{a}$ is a nonzero (integral) ideal of $\mathcal{O} = \mathcal{O}_K$, then

$$\mu(\rho(\mathfrak{a})) = \mu(\rho(\mathcal{O})) \cdot N(\mathfrak{a}) = \frac{1}{2^{r_2}}\sqrt{|d_K|}\, N(\mathfrak{a}).$$

Now choose $a > 0$ such that

$$2^{r_1} \left(\frac{\pi}{2}\right)^{r_2} \frac{a^n}{n!} = \mu(S_a(r_1, r_2))$$
$$= 2^n \mu(\rho(\mathcal{O}))N(\mathfrak{a})$$
$$= 2^n \cdot 2^{-r_2}\sqrt{|d_K|}\, N(\mathfrak{a}),$$

i.e.

$$a^n = n! \left(\frac{4}{\pi}\right)^{r_2} \sqrt{|d_K|}\, N(\mathfrak{a}).$$

By Minkowski's Lemma for (compact) convex sets, there is a nonzero element α of $\mathfrak{a}$, with $\rho(\alpha) \in S_a(r_1, r_2)$. Hence, by Theorem 5.17,

$$N(\alpha) = |\prod_\sigma \sigma(\alpha)| \le \frac{1}{n^n}\left(\sum_\sigma |\sigma(\alpha)|\right)^n \le \frac{a^n}{n^n} = \frac{n!}{n^n}\left(\frac{4}{\pi}\right)^{r_2}\sqrt{|d_K|}\, N(\mathfrak{a}).$$

[Again, here the product and the sum are taken over all $\mathbb{Q}$-isomorphisms of K into $\mathbb{C}$.] □

Corollary 5.19. *Every ideal class contains an integral ideal $\mathfrak{a}$ with*

$$N(\mathfrak{a}) \le \left(\frac{4}{\pi}\right)^{r_2} \frac{n!}{n^n}\sqrt{|d_K|}.$$

Proof. Take an ideal $\mathfrak{b}$ in the given ideal class. Multiplying by a principal ideal, we may assume that $\mathfrak{b}^{-1}$ is integral. By Theorem 5.18, choose α in $\mathfrak{b}^{-1}$ such that

$$|N(\alpha)| \le \left(\frac{4}{\pi}\right)^{r_2} \frac{n!}{n^n}\sqrt{|d_K|} \cdot N(\mathfrak{b}^{-1}). \qquad (5.12)$$

Clearly, $\mathfrak{a} = (\alpha)\mathfrak{b}$ is an integral ideal, and by (5.12),

$$N(\mathfrak{a}) = N((\alpha)\mathfrak{b}) = |N(\alpha)|N(\mathfrak{b}) \leq \left(\frac{4}{\pi}\right)^{r_2} \frac{n!}{n^n} \sqrt{|d_K|}.$$

$\square$

Definition 5.20. The constant

$$C(K) = \left(\frac{4}{\pi}\right)^{r_2} \frac{n!}{n^n}$$

is called *Minkowski's constant* for K. Note that $C(K) \to 0$ as $n \to \infty$.

Example 5.21. For $K = \mathbb{Q}(\sqrt{-5})$, $\mathcal{O}_K = \mathbb{Z}[\sqrt{-5}]$, hence

$$d_K = \begin{vmatrix} 1 & \sqrt{-5} \\ 1 & -\sqrt{-5} \end{vmatrix}^2 = -20.$$

Here $n = 2$, $r_1 = 0$, $r_2 = 1$. Therefore, Minkowski's constant is given by

$$C(K) = \frac{4}{\pi} \cdot \frac{2}{2^2} = \frac{2}{\pi}.$$

EXERCISE

Show that for all integers $n \geq 1$,

$$A_n = \left(\frac{4}{\pi}\right)^n \left(\frac{n!}{n^n}\right)^2 \geq a_n = \frac{\pi}{3}\left(\frac{3\pi}{4}\right)^{n-1}.$$

Hint: Use induction on n to show that for all $n \geq 2$, $\frac{a_n}{A_n} \leq 1$.

Theorem 5.22. *If K is a number field with $[K : \mathbb{Q}] = n > 1$, then*

$$|d_K| \geq \frac{\pi}{3}\left(\frac{3\pi}{4}\right)^{n-1}.$$

In particular, the degree $[K : \mathbb{Q}] \leq C \log d_K$, where $C > 0$ is an absolute constant.

Proof. If $\mathfrak{a} \neq (0)$ is integral, then $N(\mathfrak{a}) \geq 1$. Hence by Corollary 5.19,

$$|d_K| \geq \left(\frac{\pi}{4}\right)^{2r_2} \cdot \left(\frac{n^n}{n!}\right)^2.$$

But $\frac{\pi}{4} < 1$ and $2r_2 \leq n$. Therefore, $\left(\frac{\pi}{4}\right)^{2r_2} \geq \left(\frac{\pi}{4}\right)^n$, and hence by the Exercise above,

$$|d_K| \geq \left(\frac{\pi}{4}\right)^n \cdot \left(\frac{n^n}{n!}\right)^2 \geq \frac{\pi}{3}\left(\frac{3\pi}{4}\right)^{n-1}.$$

$\square$

Theorem 5.23 (Minkowski 1891). *Let d_K be the discriminant of the number field K. Then $d_K = 1$ if and only if $K = \mathbb{Q}$.*

Proof. It suffices to show that if $[K : \mathbb{Q}] = n > 1$, then $|d_K| > 1$. This is obvious by Theorem 5.22, because $\frac{\pi}{3} > 1$ and $\frac{3\pi}{4} > 1$. □

We need the following proposition to prove another important theorem, due to Hermite, namely, that there are only finitely many number fields of a given discriminant.

Proposition 5.24. *Given an integer $n \geq 1$ and a constant $c \geq 1$, there are only finitely many algebraic integers in $\mathbb{C}$ of degree n over $\mathbb{Q}$, such that all the conjugates of α (including α itself) are bounded by c.*

Proof. If α is an algebraic integer, then so are its conjugates. In fact, they all satisfy the same (irreducible) monic polynomial $f(x)$ in $\mathbb{Z}[x]$. Further, if $\alpha_1 = \alpha, \alpha_2, \ldots, \alpha_n$ are all the conjugates of α, then

$$
\begin{aligned}
f(x) &= x^n + a_{n-1}x^{n-1} + \cdots + a_1 x + a_0 \\
&= (x - \alpha_1) \cdots (x - \alpha_n) \\
&= x^n - (\alpha_1 + \cdots + \alpha_n)x^{n-1} + (\alpha_1\alpha_2 + \cdots)x^{n-2} - \cdots + (-1)^n \alpha_1 \ldots \alpha_n.
\end{aligned}
$$

It follows that the integer coefficients a_j of $f(x)$ are symmetric functions of $\alpha_1, \ldots, \alpha_n$ and satisfy

$$
|a_j| \leq \binom{n}{j} c^j \leq 2^n \cdot c^n = (2c)^n.
$$

This bound is independent of α and depends only on c and n. Therefore, there are only finitely many possibilities for the integer coefficients a_j of $f(x)$, and hence for α. □

EXERCISE

Suppose K is a number field with $[K : \mathbb{Q}] = n$. Let $\sigma_1, \ldots, \sigma_n$ be all the $\mathbb{Q}$-isomorphisms of K into $\mathbb{C}$. If for some α in K, $\sigma_1(\alpha) \neq \sigma_j(\alpha)$ for all $j \neq 1$, show that the degree of α over $\mathbb{Q}$ is n.

Theorem 5.25 (Hermite). *There are only finitely many number fields of a given discriminant.*

Proof. In view of Theorem 5.22, it is enough to show that there are only finitely many number fields K with fixed degree $[K : \mathbb{Q}] = n = r_1 + 2r_2$ and of discriminant $d_K = d \neq 0$. Depending on whether r_1 is zero or positive, we first define a centrally symmetric compact subset S of $\mathbb{R}^n \cong \mathbb{R}^{r_1} \times \mathbb{C}^{r_2}$, and compute its volume.

1. If $r_1 = 0$, i.e. if K is totally imaginary, S is the following product of one rectangle in $\mathbb{C}$ and $r_2 - 1$ disks, each of radius $1/2$.

$$S = \{(z_1, \ldots, z_{r_1}) \in \mathbb{C}^{r_2} |\; |\mathrm{Re}(z_1)| \leq 2^{n-2} \left(\frac{2}{\pi}\right)^{r_2-1} \sqrt{|d_K|},$$

$$|\mathrm{Im}(z_1)| \leq 1/2, |z_j| \leq 1/2 \text{ for } j = 2, \ldots, r_2\}.$$

Then

$$\mu(S) = 2 \cdot 2^{n-2} \left(\frac{2}{\pi}\right)^{r_2-1} \sqrt{|d_K|} \left(\pi \left(\frac{1}{2}\right)^2\right)^{r_2-1}$$

$$= 2^n \cdot 2^{-r_2} \sqrt{|d_K|}$$

$$= 2^n \mu(\rho\mathcal{O}).$$

2. If $r_1 > 0$, then

$$S = \{(x_1, \ldots, x_r; z_1, \ldots, z_{r_2}) \mid |x_1| \leq 2^{n-1} \left(\frac{2}{\pi}\right)^{r_2} \sqrt{|d_K|},$$

$$|x_j| \leq 1/2 \text{ for } j = 2, \ldots, r_1 \text{ and } |z_j| \leq 1/2 \text{ for } j = 1, \ldots, r_2\}.$$

Again,

$$\mu(S) = 2 \cdot 2^{n-1} \left(\frac{2}{\pi}\right)^{r_2} \sqrt{|d_K|} \left(\pi \left(\frac{1}{2}\right)^2\right)^{r_2}$$

$$= 2^n \mu(\rho\mathcal{O}).$$

Hence, by Minkowski's Lemma on (compact) convex sets, $\mathcal{O}$ contains a nonzero element α with $\rho(\alpha)$ in S.

Now

$$1 \leq |N(\alpha)| = \prod_{j=1}^{r_1} |\sigma_j(\alpha)| \prod_{j=r_1+1}^{r_1+r_2} |\sigma_j(\alpha)|,$$

and $|\sigma_j(\alpha)| \leq 1/2$ for $j = 2 \ldots, r_1 + r_2$. Therefore, $|\sigma_1(\alpha)| \geq 1$, i.e. $\sigma_1(\alpha) \neq \sigma_j(\alpha)$ for $j \neq 1$. By the Exercise above, $\deg \alpha = n$ and $K = \mathbb{Q}(\alpha)$. Thus, we have shown that each number field $K \subseteq \mathbb{C}$ with $[K : \mathbb{Q}] = n$ and $d_K = d$ is generated over $\mathbb{Q}$ by an algebraic integer α of degree n, such that all the conjugates of α are bounded by a constant $c = c(d, n)$. By Proposition 5.24, there are only finitely many such α in $\mathbb{C}$. $\square$

6

Analytic Methods

In this chapter, we shall prove Dedekind's famous formula of 1877 for the class number h_K of a number field K, namely

$$\lim_{s \to 1+} (s - 1)\zeta_K(s) = h_K \cdot \kappa, \tag{6.1}$$

where $\zeta_K(s)$ is the *Dedekind zeta function*

$$\zeta_K(s) = \sum \frac{1}{N(\mathfrak{a})^s}. \tag{6.2}$$

The summation is over all nonzero integral ideals $\mathfrak{a}$ of $\mathcal{O}_K$. We shall show that the series on the right of (6.2) converges absolutely for real s in the open interval $1 < s < \infty$. The constant κ depends only on K and can be computed explicitly. In fact, most of this chapter is devoted to the computation of κ.

Theorem 6.1. (Euler Product Formula) *For $s > 1$,*

$$\zeta_K(s) = \prod_{\mathfrak{p}} \left(1 - \frac{1}{N(\mathfrak{p})^s}\right)^{-1},$$

where the product is over all the nonzero prime ideals $\mathfrak{p}$ of K.

Proof. We only sketch the proof, leaving the details to be filled in by the reader.

First, because $N(\mathfrak{p})^s > 1$,

$$\left(1 - \frac{1}{N(\mathfrak{p})^s}\right)^{-1} = 1 + \frac{1}{N(\mathfrak{p})^s} + \frac{1}{N(\mathfrak{p})^{2s}} + \cdots$$

We formally multiply these series, one for each prime $\mathfrak{p}$, to obtain

$$\prod_{\mathfrak{p}} \left(1 - \frac{1}{N(\mathfrak{p})^s}\right)^{-1} = \sum \frac{1}{N(\mathfrak{p}_1^{e_1} \cdots \mathfrak{p}_g^{e_g})^s}.$$

In the summation, each product $\mathfrak{p}_1^{e_1} \cdots \mathfrak{p}_g^{e_g}$ occurs exactly once. Therefore, by Dedekind's unique factorization theorem for ideals,

$$\sum \frac{1}{N(\mathfrak{p}_1^{e_1} \cdots \mathfrak{p}_g^{e_g})^s} = \sum \frac{1}{N(\mathfrak{a})^s},$$

the summation being over all nonzero integral ideals of K. Hence, the Euler product formula follows.   □

6.1   Preliminaries

In Chapter 5, we proved that the group $\mathcal{O}_K^\times$ of units (of the ring of integers) of a number field K is isomorphic to $W_K \times \mathbb{Z}^r$. Here W_K is the group of roots of unity in K and $r = r_1 + r_2 - 1$. Recall that r_1 (resp. r_2) is the number of real (resp. pairs of complex) $\mathbb{Q}$-isomorphisms of K into $\mathbb{C}$, so that $[K : \mathbb{Q}] = r_1 + 2r_2$. Let $u_1, \ldots, u_r$ be a fundamental system of units in K, that is to say, any u in $\mathcal{O}_K^\times$ can be uniquely expressed as

$$u = \eta u_1^{a_1} \ldots u_r^{a_r},$$

with η in W_K and $a_1, \ldots, a_r$ in $\mathbb{Z}$. We now use the set $\{u_1, \ldots, u_r\}$ of fundamental units in K to define an important invariant of K, called its regulator, which is intimately related to its class number h_K.

The hyperplane

$$V = \{\boldsymbol{v} = (\lambda_1, \ldots, \lambda_{r_1+r_2}) \in \mathbb{R}^{r_1+r_2} | \lambda_1 + \cdots + \lambda_{r_1+r_2} = 0\}$$

is a r-dimensional subspace of $\mathbb{R}^{r_1+r_2}$. Let $\sigma_1, \ldots, \sigma_{r_1}; \sigma_{r_1+1}, \overline{\sigma}_{r_1+1}, \ldots, \sigma_{r_1+r_2}$, $\overline{\sigma}_{r_1+r_2}$ be all the $\mathbb{Q}$-isomorphisms of K into $\mathbb{C}$. In Chapter 5, we defined a map $\lambda : K^\times \to \mathbb{R}^{r_1+r_2}$ by $\lambda(\alpha) = (\log |\sigma_1(\alpha)|, \ldots, \log |\sigma_{r_1}(\alpha)|, \log |\sigma_{r_1+1}(\alpha)|^2, \ldots, \log |\sigma_{r_1+r_2}(\alpha)|^2)$, which is a group homomorphism from the multiplicative group $K^\times$ into the additive group $\mathbb{R}^{r_1+r_2}$. We proved that $\lambda(\mathcal{O}_K^\times)$ is a full lattice in the r-dimensional subspace V of $\mathbb{R}^{r_1+r_2}$, defined above. To define the regulator, we need to compute the r-dimensional volume $\mu(\lambda(\mathcal{O}_K^\times))$ of a fundamental parallelepiped of $\lambda(\mathcal{O}_K^\times)$.

6.2   The Regulator of a Number Field

We now state and prove a theorem that leads to the definition of regulator. For u in $\mathcal{O}_K^\times$, let $\lambda_j(u)$ denote the j-th component of the vector $\lambda(u)$ of $\mathbb{R}^{r_1+r_2}$.

Theorem 6.2. *The r-dimensional volume $\mu(\lambda(\mathcal{O}_K^\times))$ of any fundamental parallelepiped of the (full) lattice $\lambda(\mathcal{O}_K^\times)$ in V is given by*

$$\mu(\lambda(\mathcal{O}_K^\times)) = \sqrt{r_1 + r_2} \, |\det(\phi_i(u_j))|,$$

where $\{\phi_1, \ldots, \phi_r\}$ *is an arbitrarily chosen subset of* $\{\lambda_1, \ldots, \lambda_{r_1+r_2}\}$ *of cardinality* r.

In particular the quantity

$$R_K = |\det(\phi_i(u_j))|$$

depends only on K, *and not on the choice of* $\phi_1, \ldots, \phi_r$.

Definition 6.3. The *regulator* of a number field K is the absolute value

$$R_K = |\det(\phi_i(u_j))|$$

of the $r \times r$ determinant $\det(\phi_i(u_j))$.

Proof. Consider the unit vector

$$\boldsymbol{u} = \frac{1}{\sqrt{r_1 + r_2}} (1, \ldots, 1)$$

in $\mathbb{R}^{r_1+r_2}$. By the definition of V, the inner product $\langle \boldsymbol{u}, \boldsymbol{x} \rangle = 0$ for all $\boldsymbol{x}$ in V. Hence $\boldsymbol{u} \perp V$ ($\boldsymbol{u}$ is perpendicular to V). If

$$L = \lambda(\mathcal{O}_K^\times) \oplus \mathbb{Z}\boldsymbol{u},$$

then L is a full lattice in $\mathbb{R}^{r_1+r_2}$ and the r-dimensional volume $\mu(\lambda(\mathcal{O}_K^\times))$ of $\lambda(\mathcal{O}_K^\times)$ is equal to the (r_1+r_2)-dimensional volume of L, which is the absolute value of the $(r_1 + r_2) \times (r_1 + r_2)$ determinant

$$\frac{1}{\sqrt{r_1 + r_2}} \cdot \begin{vmatrix} 1 & \cdots & 1 \\ \lambda_1(u_1) & \cdots & \lambda_{r_1+r_2}(u_1) \\ \vdots & & \\ \lambda_1(u_r) & \vdots & \lambda_{r_1+r_2}(u_r) \end{vmatrix}.$$

For all u in $\mathcal{O}_K^\times$, we have

$$\sum_{j=1}^{r_1+r_2} \lambda_j(u) = 0.$$

Hence, if for a given m $(1 \leq m \leq r_1 + r_2)$, all other columns are added to the m-th column, and the resulting determinant is expanded by the m-th column, the above determinant is easily seen to be equal to

$$\pm \frac{1}{\sqrt{r_1 + r_2}} (r_1 + r_2) \cdot \det(\phi_i(u_j)) = \pm\sqrt{r_1 + r_2} \det(\phi_i(u_j)).$$

Since, up to a sign, the determinant does not depend on the choice of m, we are done. $\qquad\square$

6.3   Fundamental Domains

In order to compute the constant κ in the class number formula (6.1), we also need to study the so-called fundamental domain of K. Once again, recall our notation:

1. $K \subseteq \mathbb{C}$ is a number field,

2. $[K : \mathbb{Q}] = n = r_1 + 2r_2$,

3. $r = r_1 + r_2 - 1$

4. $W_K = \{\eta \in K \mid \eta^m = 1 \text{ for some } m \text{ in } \mathbb{N}\}$.

We put $w = |W_K|$. We also choose a set $u_1, \ldots, u_r$ of fundamental units of K.

Then, the set $\{\lambda(u_1), \ldots, \lambda(u_r)\}$ is a basis, over $\mathbb{R}$, of the r-dimensional subspace $V \subseteq \mathbb{R}^{r_1+r_2}$ given by

$$\lambda_1 + \cdots + \lambda_{r_1+r_2} = 0.$$

The vector $\boldsymbol{u} = \underbrace{(1, \ldots, 1; 2, \ldots, 2)}_{r_1 1s; \ r_2 2s} \notin V$. Hence, any vector $\boldsymbol{v}$ in $\mathbb{R}^{r_1+r_2}$ has a unique representation

$$\boldsymbol{v} = a_1\lambda(u_1) + \cdots + a_r\lambda(u_r) + a\boldsymbol{u},$$

with a, a_j in $\mathbb{R}$.

As before, let l be the homomorphism from the multiplicative group $\mathcal{L} = (\mathbb{R}^\times)^{r_1} \times (\mathbb{C}^\times)^{r_2}$ to the additive group $\mathbb{R}^{r_1+r_2}$, given by

$$l(x_1, \ldots, x_{r_1}; z_1, \ldots, z_{r_2}) = (\log|x_1|, \ldots, \log|x_{r_1}|; \log|z_1|^2, \ldots, \log|z_{r_2}|^2).$$
$$(6.3)$$

Definition 6.4. A set D is called a *fundamental domain* for K if D consists of the vectors $\boldsymbol{x}$ in $\mathcal{L}$, such that

1. $l(\boldsymbol{x}) = a_1\lambda(u_1) + \cdots + a_r\lambda(u_r) + a\boldsymbol{u}$ with

$0 \le a_j < 1 \ (j = 1, \ldots, r)$,

2. $0 \le \operatorname{Arg}(\boldsymbol{x}(1)) < \frac{2\pi}{w}$.

[Here $\boldsymbol{x}(1)$ stands for the first coordinate of $\boldsymbol{x}$. If $r_1 > 0$, then $\boldsymbol{x}(1) \in \mathbb{R}$ and the condition (2) means that $\boldsymbol{x}(1) > 0$.]

Example 6.5. Let us determine the fundamental domain D of a real quadratic field $K = \mathbb{Q}(\sqrt{d})$, $d > 1$ being a square-free integer. We have $[K : \mathbb{Q}] = n = 2$, $r_1 = 2, r_2 = 0$, so $r = r_1 + r_2 - 1 = 1$. If u_1 is a fundamental unit in K, then so are $\pm u_1, \pm 1/u_1$. Among them, there is one and only one which is larger than 1. Denote it by ϵ. We call ϵ the *fundamental unit* of K.

The first basis vector $\lambda(u_1)$ in equation (6.3) is $\lambda(\epsilon) = (\log |\sigma_1(\epsilon)|, \log |\sigma_2(\epsilon)|) = (\log \epsilon, -\log \epsilon)$. This is so because

$$\log |\sigma_1(\epsilon)| + \log |\sigma_2(\epsilon)| = 0.$$

Since $r_2 = 0$, $\boldsymbol{u} = (1, 1)$. Let $(x, y) \in \mathbb{R}^{r_1 + r_2} = \mathbb{R}^2$. Then equation (6.3) becomes

$$l(x, y) = (\log |x|, \log |y|) = a_1 \lambda(\epsilon) + a\boldsymbol{u} = (a_1 \log \epsilon + a, -a_1 \log \epsilon + a). \quad (6.4)$$

The fundamental domain D is determined, in this case, by the conditions

1. $xy \neq 0$,

2. $0 \le a_1 < 1$, and

3. $x > 0$.

But from (6.4), we have

$$\log |x| = a_1 \log \epsilon + a,$$

$$\log |y| = -a_1 \log \epsilon + a.$$

Subtracting, we get

$$\log \frac{|x|}{|y|} = \log \epsilon^{2a_1},$$

i.e.

$$\frac{|x|}{|y|} = \epsilon^{2a_1} \text{ or } |y| = \epsilon^{-2a_1} |x|.$$

This together with conditions (1) and (3) above imply that D consists of two components, one in the first quadrant lying between the straight lines $y = x$ and $y = \frac{1}{\epsilon^2} x$, and a similar one in the fourth quadrant. See Figure 6.1, which suggests the following.

Definition 6.6. A subset D of a real vector space is a *cone* if it contains a nonzero vector, and $c\boldsymbol{x} \in D$ whenever $\boldsymbol{x} \in D$, $c \in \mathbb{R}^+$ (the set of positive reals).

Theorem 6.7. *A fundamental domain D for a number field is a cone.*

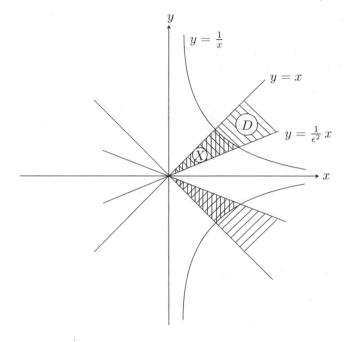

FIGURE 6.1: Fundamental domain.

Proof. Clearly, the nonzero vector $(1, \ldots, 1) \in D$. Next we show that if conditions (1) and (2) defining D hold for $\boldsymbol{x}$, then they also hold for $c\boldsymbol{x}$, given any real number $c > 0$.

(1) Let $\boldsymbol{x} = (x_1, \ldots, x_{r_1}, z_1, \ldots, z_{r_2}) \in D$, so that

$$l(\boldsymbol{x}) = a_1 \lambda(u_1) + \cdots + a_r \lambda(u_r) + a\boldsymbol{u}$$

with $0 \le a_j < 1$, $j = 1, \ldots, r$.

If $c > 0$, then

$$\begin{aligned}
l(c\boldsymbol{x}) &= (\log |cx_1|, \ldots, \log |cx_{r_1}|, \log |cz_1|^2, \ldots, \log |cz_{r_2}|^2) \\
&= l(\boldsymbol{x}) + \log(c)\boldsymbol{u} \\
&= a_1 \lambda(u_1) + \cdots + a_r \lambda(u_r) + (a + \log c)\boldsymbol{u}.
\end{aligned}$$

Hence, $c\boldsymbol{x}$ also satisfies (1).

(2) is obvious, because for a real number $c > 0$ and a complex number z, Arg $(z) =$ Arg (cz). $\square$

The reason for calling D a fundamental domain for K is that its elements form a complete set of coset representatives for the quotient group $\mathcal{L}/\rho(\mathcal{O}_K^\times)$. Recall that $D \subseteq \mathcal{L} = (\mathbb{R}^\times)^{r_1} \times (\mathbb{C}^\times)^{r_2} \subseteq \mathbb{R}^n$.

Theorem 6.8. *Let D be a domain for a number field K. Any $\boldsymbol{y}$ in $\mathcal{L}$ can be written, uniquely, as*

$$\boldsymbol{y} = \rho(u) \cdot \boldsymbol{x} \tag{6.5}$$

with u in $\mathcal{O}_K^\times$ and $\boldsymbol{x}$ in D.

[The dot in (6.5) is the component-wise multiplication in $\mathcal{L}$.]

Proof. Let η be the generator of $W_K \subseteq K$ given by

$$\eta = \cos\left(\frac{2\pi}{w}\right) + i\sin\left(\frac{2\pi}{w}\right).$$

We can certainly write

$$l(\boldsymbol{y}) = b_1\lambda(u_1) + \cdots + b_r\lambda(u_r) + b\boldsymbol{u},$$

with b, b_j in $\mathbb{R}$. Let a_j be the fractional part of b_j, i.e.

$$a_j = b_j - [b_j] \text{ with } 0 \le a_j < 1. \tag{6.6}$$

Also choose m in $\mathbb{Z}$ such that

$$0 \le \text{Arg}\,(\boldsymbol{y}(1)) - \frac{2\pi m}{w} < \frac{2\pi}{w}. \tag{6.7}$$

If we take

$$u = \eta^m u_1^{[b_1]} \cdots u_r^{[b_r]},$$

then

$$\boldsymbol{x} = \rho(u^{-1}) \cdot \boldsymbol{y} \in D.$$

In fact,

$$l(\boldsymbol{x}) = a_1\lambda(u_1) + \cdots + a_r\lambda(u_r) + b\boldsymbol{u}$$

with a_j as in (6.6), and by (6.7),

$$\text{Arg}\,(\boldsymbol{x}(1)) = \text{Arg}\,(\rho(\eta^{-m})\boldsymbol{y}(1)) = \text{Arg}\left(\boldsymbol{y}(1)) - \frac{2\pi m}{w} \in [0, 2\pi/w)\right).$$

To prove the uniqueness in (6.5), it is enough to show that if

$$\rho(u) \cdot \boldsymbol{x} = \rho(v) \cdot \boldsymbol{y}, \tag{6.8}$$

with u, v in $\mathcal{O}_K^\times$ and $\boldsymbol{x}, \boldsymbol{y}$ in D, then $u = v$ and $\boldsymbol{x} = \boldsymbol{y}$.

In equation (6.3), if $\boldsymbol{v} = \rho(v)$ with v in $\mathcal{O}_K^\times$, then all $a_j \in \mathbb{Z}$. Thus applying the map l to both sides of (6.8), and noting that two real numbers are equal if and only if their integer parts, as well as fractional parts, are equal, we get

$$l(\boldsymbol{x}) = l(\boldsymbol{y}) \text{ and thus } \lambda(u) = \lambda(v).$$

[Recall that the map $\lambda = l \circ \rho$.]

Now $\lambda(u) = \lambda(v)$ implies that $v = \omega u$ for ω in W_K. Hence by (6.8),

$$\boldsymbol{x} = \rho(\omega)\boldsymbol{y},$$

which implies that

$$\mathrm{Arg}\ (\boldsymbol{x}(1)) =\ \mathrm{Arg}\ (\boldsymbol{y}(1)) +\ \mathrm{Arg}\ (\omega). \tag{6.9}$$

Since

$$0 \leq\ \mathrm{Arg}\ (\boldsymbol{x}(1)),\ \mathrm{Arg}\ (\boldsymbol{y}(1)) < \frac{2\pi}{w}$$

and $\mathrm{Arg}\ (\omega)$ is a multiple of $\frac{2\pi}{w}$, equation (6.9) is possible only if $\mathrm{Arg}\ (\omega) = 0$, which implies that $\omega = 1$. Hence, $u = v$, which also gives $\boldsymbol{x} = \boldsymbol{y}$. $\qquad\square$

Definition 6.9. Let $\alpha, \beta \in \mathcal{O}_K^\times$. We say that α and β are *associates* if $\beta = u\alpha$ with some u in $\mathcal{O}_K^\times$.

Being associate is an equivalence relation, and hence partitions the nonzero elements of $\mathcal{O}_K$ into the set $\mathfrak{C}$ of equivalence classes. Let D be a fundamental domain for K. Theorem 6.8 defines an injective map

$$f : \mathfrak{C} \to D$$

as follows. For $\alpha \neq 0$ in $\mathcal{O}_K$, write (uniquely) $\rho(\alpha) = \rho(u) \cdot \boldsymbol{x}$ with $\boldsymbol{x}$ in D and u in $\mathcal{O}_K^\times$. Put $f(\alpha) = \boldsymbol{x}$. Clearly if α, β are associates, then $f(\alpha) = f(\beta)$. Therefore, for C in $\mathfrak{C}$, we put

$$f(C) = f(\alpha) = \boldsymbol{x} \tag{6.10}$$

for any α in C.

Recall defining the *norm function* $N\ :\ \mathcal{L}\ \to\ \mathbb{R}$. For $\boldsymbol{x}\ =\ (x_1, \ldots, x_{r_1}, z_1, \ldots, z_{r_2})$, $N(\boldsymbol{x}) = x_1 \cdots x_{r_1} |z_1|^2 \cdots |z_{r_2}|^2$.

Theorem 6.10. *Suppose $D \subseteq \mathcal{L}$ is a fundamental domain for K. The restricted fundamental domain, that is, the set*

$$X = X_D = \{\boldsymbol{x} \in D \,|\, N(\boldsymbol{x}) \leq 1\}$$

is measurable with measure

$$\mu(X) = 2^{r_1} \pi^{r_2} R_K / w. \tag{6.11}$$

We break the proof into simpler parts.

Lemma 6.11. *Let* $c = (c_1, \ldots, c_{r_1}, s_1, \ldots, s_{r_2}) \in \mathcal{L}$. *If* $S \subseteq \mathcal{L}$ *is a (Lebesgue) measurable set, then so is* $c \cdot S$ *and*

$$\mu(c \cdot S) = |N(c)|\mu(S).$$

In particular, if $|N(c)| = 1$, *the measure* μ *is invariant under the map*

$$\mathcal{L} \ni x \to c \cdot x \in \mathcal{L}.$$

Proof. If S is a measurable subset of $\mathbb{R}$ and $c \in \mathbb{R}$, then $\mu(cS) = |c|\mu(S)$.

Now let $s = \rho_0 e^{i\theta_0} \in \mathbb{C}$. Under the multiplication by s, i.e.

$$z = \rho e^{i\theta} \to \rho_0 \rho e^{i(\theta + \theta_0)},$$

$$\begin{aligned} d\mu(z) = \rho d\rho d\theta &\to \rho \rho_0 d(\rho \rho_0) d(\theta + \theta_0) \\ &= \rho_0^2 \rho d\rho d\theta \\ &= |s|^2 \cdot d\mu(z). \end{aligned}$$

The Lebesgue measure on $\mathcal{L} = (\mathbb{R}^\times)^{r_1} \times (\mathbb{C}^\times)^{r_2}$ is the product of the Lebesgue measures on its factors. Therefore, if $S \subseteq \mathcal{L}$ is measurable, then

$$\begin{aligned} \mu(c \cdot S) &= |c_1| \cdots |c_{r_1}||s_1|^2 \cdots |s_{r_2}|^2 \mu(S) \\ &= |N(c)|\,\mu(S). \end{aligned}$$

$\square$

Lemma 6.12. *The restricted fundamental domain*

$$X = \{x \in D \mid N(x) \leq 1\}$$

is bounded.

Proof. First take $x = (x_1, \ldots, x_{r_1}, z_1, \ldots, z_{r_2}) \in X = X_D$ with $N(x) = 1$. Write

$$l(x) = a_1 \lambda(u_1) + \cdots + a_r \lambda(u_r) + a u \tag{6.12}$$

with

$$0 \leq a_j < 1, \; j = 1, \ldots, r.$$

Since for a unit $u \in \mathcal{O}_K^\times$, $\sum_{j=1}^r \sigma_j(u) = 0$, adding the coordinates on each side of the vector equation (6.12), we get

$$\begin{aligned} \log|N(x)| &= \log|x_1| + \cdots + \log|x_{r_1}| + \log|z_1|^2 + \cdots + \log|z_{r_2}|^2 \\ &= \sum_{i=1}^r a_i (\log|\sigma_1(u_i)| + \cdots + \log|\sigma_{r_1}(u_i)| + \log|\sigma_{r_1+1}(u_i)|^2 + \cdots \\ &\quad + \log|\sigma_{r_1+r_2}(u_i)|^2) + a(1 + \cdots + 1 + 2 + \cdots + 2) \\ &= an. \end{aligned}$$

Hence

$$a = \frac{\log |N(\boldsymbol{x})|}{n}. \tag{6.13}$$

Since $|N(\boldsymbol{x})| = 1$, we get $a = 0$. Therefore, the subset $T = \{\boldsymbol{x} \in X \mid |N(\boldsymbol{x})| = 1\}$ of X is bounded, because it consists of $\boldsymbol{x}$ in X with

$$l(\boldsymbol{x}) = a_1 \lambda(u_1) + \cdots + a_r \lambda(u_r),$$

where

$$0 \leq a_j < 1.$$

The set X can be characterized by

$$X = \{c\boldsymbol{x} \mid \boldsymbol{x} \in T, \ 0 < c \leq 1\}.$$

Hence X is also bounded. $\qquad \square$

Lemma 6.13. *Let Y be the set $\{\boldsymbol{y} \in \mathcal{L} \mid N(\boldsymbol{y}) \leq 1, \ l(\boldsymbol{y}) = a_1 \lambda(u_1) + \cdots + a_r \lambda(u_r) + a\boldsymbol{u} \text{ with } 0 \leq a_j < 1\}$, with no restriction on $\mathrm{Arg}\, (\boldsymbol{y}(1))$. Then*

$$\mu(Y) = w\mu(X).$$

Proof. Let

$$\eta = \cos \frac{2\pi}{w} + i \sin \frac{2\pi}{w},$$

so that η is a generator of the group W_K of roots of unity in K. For each j $(0 \leq j < w)$, we put

$$X_j = \rho(\eta^j) \cdot X.$$

If $\boldsymbol{x}$ is in X, then

1. $|N(\rho(\eta^j) \cdot \boldsymbol{x})| = |N(\rho(\eta^j))||N(\boldsymbol{x})| = |N(\boldsymbol{x})|,$
2. $l(\rho(\eta^j) \cdot \boldsymbol{x}) = l(\rho(\eta^j)) + l(\boldsymbol{x}) = l(\boldsymbol{x}),$
3. $\mathrm{Arg}\, (\sigma_1(\eta^j)\boldsymbol{x}(1)) = \mathrm{Arg}\, (\boldsymbol{x}(1)) + \frac{2\pi}{w} j.$

Therefore, Y is a disjoint union of $X_1, \ldots, X_w$ ($X_w = X_0$). By Lemma 6.11, $\mu(X_j) = \mu(X)$, for all $j = 1, \ldots, w$. This proves that $\mu(Y) = w\mu(X)$. $\quad \square$

Proposition 6.14. *The set*

$$Y^+ = \{(x_1, \ldots, x_{r_1}, z_1, \ldots, z_{r_2}) \in Y \mid x_j > 0, \ \forall \ j\}$$

is measurable and

$$\mu(Y) = 2^{r_1} \mu(Y^+).$$

Proof. The 2^{r_1} vectors $s = (\underbrace{\pm 1, \ldots, \pm 1}_{r_1 \text{ components}}; \underbrace{1, \ldots, 1}_{r_2 \text{ components}})$ partition Y into a disjoint union of 2^{r_1} sets $s \cdot Y^+$, each of measure $\mu(Y^+)$. (Lemma 6.11, since $|N(s)| = 1$.) Therefore,

$$\mu(Y) = 2^{r_1} \mu(Y^+). \qquad \square$$

Corollary 6.15.

$$\mu(X) = \frac{2^{r_1} \mu(Y^+)}{w}.$$

Proof. Combine Proposition 6.14 and Lemma 6.13. $\qquad \square$

Proposition 6.16. $\mu(Y^+) = \pi^{r_2} R_K$.

Proof. Recall that Y^+ consists of the vectors $\boldsymbol{x} = (x_1, \ldots, x_{r_1}, z_1, \ldots, z_{r_2})$ in $\mathcal{L}$, such that

1. $x_j > 0, \forall j = 1, \ldots, r_j$;

2. $N(\boldsymbol{x}) \leq 1$ (Note that (1) implies $|N(\boldsymbol{x})| = N(\boldsymbol{x})$.);

3. if

$$l(\boldsymbol{x}) = \tau_1 \lambda(u_1) + \cdots + \tau_r \lambda(u_r) + \frac{\log |N(\boldsymbol{x})|}{n} \boldsymbol{u}, \qquad (6.14)$$

   then $0 \leq \tau_j < 1$ for all $j = 1, \ldots, r_1$.

Let

$$z_j = s_j + \sqrt{-1}\, t_j, \; j = 1, \ldots, r_2.$$

We change the variables

$$s_j = \rho_j \cos \theta_j,$$
$$t_j = \rho_j \sin \theta_j$$

with $\rho_j > 0$ and $\theta_j \in [0, 2\pi)$, so that

$$\mu(Y^+) = \int_{Y^+} dx_1 \ldots dx_{r_1} \rho_1 \ldots \rho_{r_2} d\rho_1 \ldots d\rho_{r_2} d\theta_1 \ldots d\theta_{r_2},$$

i.e.

$$\mu(Y^+) = (2\pi)^{r_2} \int \ldots \int \rho_1 \ldots \rho_{r_2} d\rho_1 \ldots d\rho_{r_2} dx_1 \ldots dx_{r_1}. \qquad (6.15)$$

We use equation (6.14) to change variables $x_1, \ldots, x_{r_1}; \rho_1, \ldots, \rho_{r_2}$ (which satisfy

1. $x_j > 0$ for all $j = 1, \ldots, r_1$;

  2. $(x_1 \ldots x_{r_1})(\rho_1^2 \ldots \rho_{r_2}^2) \le 1$

to $\tau_1, \ldots, \tau_r, \tau = N(\boldsymbol{x})$ (satisfying $0 < \tau_j \le 1$, $j = 1, \ldots, r_1$; and $0 < \tau \le 1$).

To compute the Jacobian of this substitution, we rewrite equation (6.14) as

$$\log x_i = \sum_{j=1}^{r} \tau_j \lambda_i(u_j) + \frac{\log \tau}{n}, \ i = 1, \ldots, r_1, \tag{6.16}$$

and

$$2 \log \rho_i = \sum_{j=1}^{r} \tau_j \lambda_i(u_j) + \frac{2 \log \tau}{n}, \ i = 1, \ldots, r_2. \tag{6.17}$$

For $i = 1, \ldots, r_1$, equation (6.16) gives

$$\frac{\partial x_i}{\partial \tau_j} = x_i \lambda_i(u_j), \ j = 1, \ldots, r.$$

For $i = 1, \ldots, r_2$, equation (6.17) gives

$$\frac{\partial \rho_i}{\partial \tau_j} = \frac{\rho_i}{2} \lambda_i(u_j), \ j = 1, \ldots, r.$$

Finally,

$$\frac{\partial x_i}{\partial \tau} = \frac{x_i}{n\tau}, \ i = 1, \ldots, r_1.$$

and

$$\frac{\partial \rho_i}{\partial \tau} = \frac{\rho_i}{n\tau}, \ i = 1, \ldots, r_2.$$

Hence the Jacobian J is the absolute value of the determinant

$$\begin{vmatrix} x_1 \lambda_1(u_1) & \cdots & x_1 \lambda_1(u_r) & x_1/n\tau \\ \vdots & & & \\ x_{r_1} \lambda_{r_1}(u_1) & \cdots & x_{r_1} \lambda_{r_1}(u_r) & x_{r_1}/n\tau \\ \rho_1 \lambda_{r_1+1}(u_1)/2 & \cdots & \rho_1 \lambda_{r_1+1}(u_r)/2 & \rho_1/n\tau \\ \vdots & & & \\ \rho_{r_2} \lambda_{r_1+r_2}(u_1)/2 & \cdots & \rho_{r_2} \lambda_{r_1+r_2}(u_r)/2 & \rho_{r_2}/n\tau \end{vmatrix}$$

$$= \frac{x_1 \cdots x_{r_1} \rho_1 \cdots \rho_{r_2}}{n 2^{r_2} \tau} \begin{vmatrix} \lambda_1(u_1) & \cdots & \lambda_1(u_r) & 1 \\ \vdots & & & \\ \lambda_{r_1}(u_1) & \cdots & \lambda_{r_1}(u_r) & 1 \\ \lambda_{r_1+1}(u_1) & \cdots & \lambda_{r_1+1}(u_r) & 2 \\ \vdots & & & \\ \lambda_{r_1+r_2}(u_1) & \cdots & \lambda_{r_1+r_2}(u_r) & 2 \end{vmatrix}$$

If we add other rows to the first row, and then note that

1. $\sum_{j=1}^{r_1+r_2} \lambda_j(u) = 0, \forall\, u \in \mathcal{O}_K^\times$, and

2. $\tau = x_1 \cdots x_{r_1} \rho_1^2 \cdots \rho_{r_2}^2$,

we obtain

$$J = \frac{R_K}{2^{r_2} \rho_1 \cdots \rho_{r_2}}.$$

Hence, (6.15) becomes

$$\mu(Y^+) = (2\pi)^{r_2} \cdot \int_{Y^+} \rho_1 \cdots \rho_{r_2} \cdot J \cdot d\rho_1 \cdots d\rho_{r_2} dx_1 \cdots dx_{r_1}$$

$$= \pi^{r_2} R_K \int_0^1 \cdots \int_0^1 d\rho_1 \cdots d\rho_{r_2} dx_1 \cdots dx_{r_1}$$

$$= \pi^{r_2} R_K. \qquad \square$$

Proof of Theorem 6.10. By Corollary 6.15 and Proposition 6.16,

$$\mu(X) = \frac{2^{r_1} \mu(Y^+)}{w}$$

$$= \frac{2^{r_1} \pi^{r_2} R_K}{w}. \qquad \square$$

6.4 Zeta Functions

We now return to the Dedekind class number formula.

6.4.1 The Riemann Zeta Function

The most famous zeta function is the *Riemann zeta* function $\zeta(s)$ defined for $s = \sigma + it$ in $\mathbb{C}$ with $\sigma > 1$ by

$$\zeta(s) = \sum_{m=1}^{\infty} \frac{1}{m^s}. \tag{6.18}$$

However, throughout this chapter, we shall assume that $t = 0$, that is $s \in \mathbb{R}$.

Theorem 6.17. *The series for $\zeta(s)$ in* (6.18) *converges for $s > 1$ and*

$$\lim_{s \to 1+} (s-1)\zeta(s) = 1. \tag{6.19}$$

Proof. Let $s > 1$. For $x \in (1, \infty)$, $\frac{1}{x^s}$ is a decreasing function. Hence,

$$\int_m^{m+1} \frac{dx}{x^s} < \frac{1}{m^s} < \int_{m-1}^m \frac{dx}{x^s}.$$

Therefore, for $N > 2$,

$$\int_1^N \frac{dx}{x^s} < \sum_{m=1}^N \frac{1}{m^s} < 1 + \int_1^N \frac{dx}{x^s},$$

which gives

$$\int_1^\infty \frac{dx}{x^s} < \zeta(s) < 1 + \int_1^\infty \frac{dx}{x^s},$$

i.e.

$$\frac{1}{s-1} < \zeta(s) < 1 + \frac{1}{s-1}.$$

Multiply this inequality throughout by $s - 1$ and let $s \to 1+$, to obtain (6.19).

$\square$

6.4.2   A Partial Zeta Function

Suppose D is a fundamental domain for a number field K with degree $[K : \mathbb{Q}] = n = r_1 + 2r_2$. Let

$$X = X_D = \{\boldsymbol{x} \in D | N(\boldsymbol{x}) \leq 1\}$$

be the restricted fundamental domain. For t in $\mathbb{R}$, $N(t\boldsymbol{x}) = t^n N(\boldsymbol{x})$. Let L be a lattice in $\mathbb{R}^{r_1} \times \mathbb{C}^{r_2} \cong \mathbb{R}^n$. For a real number s, define the *partial zeta function* $Z(s) = Z(L, D, s)$ by

$$Z(s) = \sum_{\boldsymbol{x} \in L \cap D} \frac{1}{|N(\boldsymbol{x})|^s}. \tag{6.20}$$

Clearly, $Z(s)$ depends on L and D.

Theorem 6.18. *The series for $Z(s)$ on the right of (6.20) converges for $s > 1$ and*

$$\lim_{s \to 1+} (s - 1)Z(s) = \mu(X)/\mu(L).$$

Proof. For $t \in \mathbb{R}$, $t > 0$ and $S \subseteq \mathbb{R}^n$, let

$$tS = \{t\boldsymbol{x} \,|\, \boldsymbol{x} \in S\}.$$

Since L is discrete and X is bounded, the number

$$\nu(t) = |tX \cap L| = \left| X \cap \frac{1}{t} L \right|$$

of points common to both tX and L is finite. Moreover, if $\Delta = \mu(L)$, then

$$v := \mu(X) = \lim_{t \to \infty} \Delta \frac{\nu(t)}{t^n}. \tag{6.21}$$

This is so, because $\frac{1}{t} L$ provides a disjoint cover of X by $\nu(t)$ parallelepipeds, each of measure Δ/t^n, for an upper Riemann sum to approximate $\mu(X)$ from above. As t goes to infinity, we get $\mu(X)$.

Since $L \subseteq \mathbb{R}^{r_1} \times \mathbb{C}^{r_2} \cong \mathbb{R}^n$ is discrete and the norm on L is a non-constant continuous function, there are only finitely many points in L of bounded norm. Hence, we can arrange the points of $L \cap D$ in a sequence $\{\boldsymbol{x}_m\}$, such that

$$1 \le |N(\boldsymbol{x}_1)| \le |N(\boldsymbol{x}_2)| \le |N(\boldsymbol{x}_3)| \le \cdots .$$

Let

$$t_m = |N(\boldsymbol{x}_m)|^{1/n}.$$

Because

$$\begin{aligned} \nu(t_m) &= |L \cap t_m X| \\ &= \{\gamma \in L \cap D \,|\, |N(\gamma)| \le t_m\}, \end{aligned}$$

we have

$$\nu(t_m) \ge m. \tag{6.22}$$

On the other hand, for any $\epsilon > 0$, $\boldsymbol{x}_j \notin (t_m - \epsilon)X$, if $j \ge m$. Hence,

$$\nu(t_m - \epsilon) < m \le \nu(t_m). \tag{6.23}$$

Because $t_m^n = |N(\boldsymbol{x}_m)|$, we obtain

$$\frac{\nu(t_m - \epsilon)}{t_m^n} < \frac{m}{|N(\boldsymbol{x}_m)|} \le \frac{\nu(t_m)}{t_m^n}. \tag{6.24}$$

Finally, (6.21) and (6.24) imply that

$$\lim_{m \to \infty} \frac{m}{|N(\boldsymbol{x}_m)|} = \frac{v}{\Delta}. \tag{6.25}$$

The equality (6.25) can now be used to compare the series in (6.18) and (6.20), i.e. to say that for $s > 1$, $\sum_{m=1}^{\infty} \frac{1}{m^s}$ converges if and only if

$$\sum_{\boldsymbol{x} \in L \cap D} \frac{1}{|N(\boldsymbol{x})|^s} = \sum \frac{1}{|N(\boldsymbol{x}_m)|^s}$$

does.

Again, if $s > 1$ and $\epsilon > 0$, by (6.25), for all sufficiently large m, say $m > m_0$ for some m_0, we have

$$\left(\frac{v}{\Delta} - \epsilon\right) < \frac{m}{|N(\boldsymbol{x}_m)|} < \frac{v}{\Delta} + \epsilon.$$

This gives

$$\left(\frac{v}{\Delta} - \epsilon\right)^s \frac{1}{m^s} < \frac{1}{|N(\boldsymbol{x}_m)|^s} < \left(\frac{v}{\Delta} + \epsilon\right)^s \frac{1}{m^s}.$$

Hence,

$$\left(\frac{v}{\Delta} - \epsilon\right)^s \sum_{m=m_0}^{\infty} \frac{1}{m^s} < \sum_{m=m_0+1}^{\infty} \frac{1}{|N(\boldsymbol{x}_m)|^s} < \left(\frac{v}{\Delta} + \epsilon\right)^s \sum_{m=m_0+1}^{\infty} \frac{1}{m^s}. \quad (6.26)$$

But

$$\lim_{s\to1+} (s-1) \sum_{m=1}^{m_0} \frac{1}{|N(\boldsymbol{x}_m)|^s} = \lim_{s\to1+} (s-1) \sum_{m=1}^{m_0} \frac{1}{m^s} = 0. \quad (6.27)$$

Therefore, multiplying (6.26) throughout by $s-1$, taking the limit as $s \to 1+$, and adding (6.27) to the limit, by Theorem 6.17, we get

$$\frac{v}{\Delta} - \epsilon \leq \lim_{s\to1+} Z(s) \leq \frac{v}{\Delta} + \epsilon.$$

Since $\epsilon > 0$ can be arbitrarily small, this proves Theorem 6.18.   □

6.4.3   The Dedekind Zeta Function

Let K be a number field of degree $[K : \mathbb{Q}] = n = r_1 + 2r_2$. Recall that for $s = \sigma + it$ in $\mathbb{C}$, $\sigma > 1$, the Dedekind zeta function $\zeta_K(s)$ of K is defined by

$$\zeta_K(s) = \sum_{\mathfrak{a}} \frac{1}{N(\mathfrak{a})^s}, \quad (6.28)$$

where the summation is over all nonzero integral ideals $\mathfrak{a}$ of $\mathcal{O}_K$. In particular, if $K = \mathbb{Q}$, all the integral ideals $\mathfrak{a}$ are of the form $\mathfrak{a} = n\mathbb{Z}$ for n in $\mathbb{N}$, and $N(\mathfrak{a}) = n$. Hence the Dedekind zeta function

$$\zeta_{\mathbb{Q}}(s) = \sum_{m=1}^{\infty} \frac{1}{m^s}$$

is just the Riemann zeta fuction $\zeta(s)$.

Let $h = h_K$ be the class number of K and $\{C_1, \ldots, C_h\}$ be its ideal class group. We write (6.28) as

$$\zeta_K(s) = \sum_{j=1}^{h} \zeta_{C_j}(s),$$

where

$$\zeta_{C_j}(s) = \sum_{\mathfrak{b}} \frac{1}{N(\mathfrak{b})^s},$$

the summation being over all integral ideals $\mathfrak{b}$ in C_j.

We will restrict s to be in $\mathbb{R}$, and show that

1. each $\zeta_{C_j}(s)$ converges for $s > 1$ and

2. $\lim_{s \to 1+}(s-1)\zeta_{C_j}(s)$ exists, and is independent of $j = 1, \ldots, h$.

Let C be one of $C_1, \ldots, C_h$ and choose an integral ideal $\mathfrak{a}$ in C^{-1}, so that for every integral ideal $\mathfrak{b}$ in C, $\mathfrak{ab} = (\alpha) = \alpha \mathcal{O}_K$. Since this means that $\alpha \in \mathfrak{a}$, we get a bijective map $\mathfrak{b} \to (\alpha)$, α in $\mathfrak{a}$, mapping integral ideals $\mathfrak{b}$ in C to the principal ideals generated by nonzero elements of $\mathfrak{a}$. Moreover, $|N_{K/\mathbb{Q}}(\alpha)| = N(\mathfrak{ab}) = N(\mathfrak{a})N(\mathfrak{b})$. Thus we have

$$N(\mathfrak{b}) = |N_{K/\mathbb{Q}}(\alpha)|/N(\mathfrak{a}),$$

and

$$\zeta_C(s) = N(\mathfrak{a})^s \sum_{\substack{(\alpha) \\ \alpha \in \mathfrak{a}}} \frac{1}{|N_{K/\mathbb{Q}}(\alpha)|^s}.$$

Now $(\alpha) = (\beta)$ if and only if α and β are associates $\Leftrightarrow |N(\alpha)| = |N(\beta)|$. (We write $N_{K/\mathbb{Q}}(\alpha)$ simply as $N(\alpha)$.) Therefore

$$\sum_{\substack{(\alpha) \\ \alpha \in \mathfrak{a}}} \frac{1}{|N(\alpha)|^s} = \sum \frac{1}{|N(\alpha)|^s},$$

the last sum being over all the pairwise non-associate elements $\alpha \neq 0$ of $\mathfrak{a}$.

Let D be a fundamental domain of K. Equation (6.10) defines a bijective map $\mathfrak{C} \ni \alpha \to f(\alpha) = \boldsymbol{x} \in D$, from the set of equivalence classes of non-associate elements of $\mathcal{O}_K$ to D. Therefore, because $N(\alpha) = N(f(\alpha))$,

$$\sum_{\substack{(\alpha) \\ \alpha \in \mathfrak{a}}} \frac{1}{|N(\alpha)|^s} = \sum_{\boldsymbol{x} \in L \cap D} \frac{1}{|N(\boldsymbol{x})|^s} = Z(L, D, s),$$

with the lattice $L = \rho(\mathfrak{a})$. This shows that the series for $\zeta_C(s)$ converges for $s > 1$. Further, by Theorem 6.18,

$$\lim_{s \to 1+}(s-1)\zeta_C(s) = \lim_{s \to 1+} N(\mathfrak{a})^s(s-1)Z(L, D, s)$$

$$= \lim_{s \to 1+} N(\mathfrak{a})^s \frac{\mu(X)}{\mu(L)},$$

$$= \lim_{s \to 1+} N(\mathfrak{a})^s \cdot \frac{2^{r_1}\pi^{r_2}R_K}{w} \cdot \frac{1}{2^{-r_2}\sqrt{|d_K|}N(\mathfrak{a})}$$

$$= \frac{2^{r_1+r_2}\pi^{r_2}R_K}{w\sqrt{|d_K|}}$$

$$= \kappa,$$

proving that this limit κ is independent of the ideal class C. Hence, summing over all C,

$$\lim_{s \to 1+} (s-1)\zeta_K(s) = \kappa \cdot h_K.$$

This is the Dedekind class number formula we set out to prove.

Theorem 6.19 (Dedekind). *We have*

$$\lim_{s \to 1+} (s-1)\zeta_K(s) = \frac{2^{r_1+r_2} \pi^{r_2} R_K}{w\sqrt{|d_K|}} \cdot h_K.$$

EXERCISE

Specialize Theorem 6.19 when K is a quadratic field.

7

Arithmetic in Galois Extensions

In 1894, Hilbert developed decomposition theory to determine the factorization of a prime ideal $\mathfrak{p}$ of k in K for Galois extensions K/k in terms of the Galois group $\operatorname{Gal}(K/k)$.

An extension K/k of number fields is a *Galois extension* if it is a *normal extension*, i.e. for every k-isomorphism $\sigma : K \to \mathbb{C}$, we have $\sigma(K) \subseteq K$. [A ring homomorphism $\sigma : K \to K$ is a *k-isomorphism* if its restriction to k is the identity map.] Some important examples of Galois extensions are quadratic extensions, and the cyclotomic extensions $\mathbb{Q}(\zeta)$, where ζ is a root of unity. Throughout this chapter, we assume that K/k is a Galois extension of degree n. The set $G = \operatorname{Gal}(K/k)$ of n $\mathbb{Q}$-isomorphisms $\sigma_1, \ldots, \sigma_n$ of K into K is a group under the composition of maps. We call $\operatorname{Gal}(K/k)$ the *Galois group* of K over k.

Again let $\mathcal{O} = \mathcal{O}_K$, $\mathfrak{o} = \mathcal{O}_k$ and write for a prime $\mathfrak{p}$ of k,

$$\mathfrak{p}\mathcal{O} = \mathfrak{P}_1^{e_1} \cdots \mathfrak{P}_g^{e_g}. \tag{7.1}$$

If $\sigma \in G$ and $\mathfrak{P}$ is a prime in K, then so is $\sigma(\mathfrak{P})$. Since $\sigma(\mathfrak{p}\mathcal{O}) = \mathfrak{p}\mathcal{O}$, from (7.1) we get

$$\mathfrak{p}\mathcal{O} = \sigma(\mathfrak{P}_1)^{e_1} \cdots \sigma(\mathfrak{P}_g)^{e_g}. \tag{7.2}$$

By the uniqueness of factorization, it follows from (7.1) and (7.2) that given i, $\sigma(\mathfrak{P}_i) = \mathfrak{P}_j$ for some j. Conversely, we prove that given j, $\mathfrak{P}_j = \sigma(\mathfrak{P}_i)$ for some i.

Theorem 7.1. *If $\mathfrak{P}$ and $\mathfrak{Q}$ are two prime ideals of K dividing a prime $\mathfrak{p}$ in k, then $\mathfrak{Q} = \sigma(\mathfrak{P})$ for some σ in the Galois group $G = \operatorname{Gal}(K/k)$.*

Proof. Suppose not, i.e. $\sigma(\mathfrak{P}) \neq \mathfrak{Q}$ for all σ in G. By the Chinese Remainder Theorem, we may choose α in $\mathcal{O}$, such that

$$\alpha \equiv 0 \pmod{\mathfrak{Q}}$$

and

$$\alpha \equiv 1 \pmod{\sigma(\mathfrak{P})}$$

for all σ in G. If $\sigma_1 = 1_K$, then $N_{K/k}(\alpha) = \alpha \cdot \sigma_2(\alpha) \cdots \sigma_n(\alpha) \in \mathfrak{Q} \cap \mathfrak{o} = \mathfrak{p}$. But $\alpha \equiv 1 \pmod{\sigma(\mathfrak{P})}$ implies that $\sigma(\alpha) \notin \mathfrak{Q}$, for all σ in G. Hence $N_{K/k}(\alpha) = \prod_{\sigma \in G} \sigma(\alpha) \notin \mathfrak{Q} \supseteq \mathfrak{p}$. This is a contradiction. $\square$

Corollary 7.2. *Suppose K/k is a Galois extension of degree n, $\mathfrak{p}$ a prime of k and*

$$\mathfrak{p}\mathcal{O} = \mathfrak{P}_1^{e_1} \cdots \mathfrak{P}_g^{e_g}.$$

Then $e_1 = \ldots = e_g = e$, say, and $f_1 = \ldots = f_g = f$, say. Hence $n = efg$.

Proof. Recall Definition 4.5 of f. Given i, j ($1 \leq i < j \leq g$), the primes $\mathfrak{P}_i$, $\mathfrak{P}_j$ in the factorization of (7.1) of $\mathfrak{p}$ are interchanged by some σ in G, but not e_i, e_j. By the uniqueness of factorization, it follows that then $e_i = e_j$. If $\sigma(\mathfrak{P}_i) = \mathfrak{P}_j$, then σ induces an isomorphism $\mathcal{O}/\mathfrak{P}_i \cong \mathcal{O}/\mathfrak{P}_j$. This shows that $f_i = f_j$. Therefore $n = e_1 f_1 + \cdots + e_g f_g = efg$. $\qquad\square$

7.1   Hilbert Theory

Let K/k be a Galois extension of number fields with $G = \mathrm{Gal}(K/k)$. To study the factorization (7.1), it is enough, by Corollary 7.2, to fix a prime $\mathfrak{p}$ of k and prime $\mathfrak{P}$ of K dividing $\mathfrak{p}$.

Definition 7.3. The *decomposition group* Z of $\mathfrak{P}/\mathfrak{p}$ is defined to be

$$Z = Z_{\mathfrak{P}/\mathfrak{p}} = \{\sigma \in G \,|\, \sigma(\mathfrak{P}) = \mathfrak{P}\}.$$

The *inertia group* $T = T_{\mathfrak{P}/\mathfrak{p}}$ of $\mathfrak{P}/\mathfrak{p}$ is the set

$$T = \{\sigma \in G \,|\, \sigma(\alpha) \equiv \alpha \pmod{\mathfrak{P}}, \forall \alpha \in \mathcal{O}\}.$$

It is easy to see that Z and T are subgroups of G. Moreover, for α in $\mathfrak{P}$, the condition $\sigma(\alpha) \equiv \alpha \pmod{\mathfrak{P}}$ becomes $\sigma(\alpha) \equiv 0 \pmod{\mathfrak{P}}$. Hence $\sigma(\mathfrak{P}) = \mathfrak{P}$. Therefore, T is a subgroup of Z. The letter Z and T are traditional (for their German equivalents, Zerlegung for decomposition and Trägheit for inertia).

For every σ in G, $\sigma(\mathcal{O}) = \mathcal{O}$. Hence, if further $\sigma \in Z$, i.e. $\sigma(\mathfrak{P}) = \mathfrak{P}$, then we have the obvious map

$$\bar{\sigma} : \mathcal{O}/\mathfrak{P} \to \mathcal{O}/\mathfrak{P}$$

taking the coset $\alpha + \mathfrak{P}$ to the coset $\sigma(\alpha) + \mathfrak{P}$. Because $\sigma|_{\mathfrak{o}} = 1_{\mathfrak{o}}$, it is clear that $\sigma|_{\mathfrak{o}/\mathfrak{p}} = 1_{\mathfrak{o}/\mathfrak{p}}$. Hence if we put $\overline{K} = \mathcal{O}/\mathfrak{P}$ and $\overline{k} = \mathfrak{o}/\mathfrak{p}$, then $\bar{\sigma} \in \mathrm{Gal}(\overline{K}/\overline{k})$. [Recall that any extension of finite fields is Galois.] Thus, we have proved the following fact.

Theorem 7.4. *The map*

$$Z \ni \sigma \to \bar{\sigma} \in \mathrm{Gal}(\overline{K}/\overline{k})$$

is a group homomorphism with kernel T. Hence T is a normal subgroup of Z and Z/T is isomorphic to a subgroup of $\mathrm{Gal}(\overline{K}/\overline{k})$. In particular, the order $|Z/T|$ divides the order $f = |\mathrm{Gal}\,\overline{K}/\overline{k}|$.

For a subgroup H of G and a subset X of K, let X^H be the subset of X defined by

$$X^H = \{x \in X \,|\, \sigma(x) = x \text{ for all } \sigma \text{ in } H\}.$$

Thus $K^G = k$, $K^{\{e\}} = K$, $\mathcal{O}^G = \mathfrak{o}$ and for a prime $\mathfrak{P}$ in K dividing a prime $\mathfrak{p}$ in k, $\mathfrak{P}^G = \mathfrak{p}$. In general, K^H is an intermediate field between k and K, called the *fixed field* of H. It is easy to see that $\mathcal{O}_{K^H} = \mathcal{O}_K^H$ and $\mathfrak{P}^H = \mathfrak{P} \cap \mathcal{O}^H$. Moreover, the residue field $\mathcal{O}^H / \mathfrak{P}^H$ is an intermediate field between $\mathfrak{o}/\mathfrak{p}$ and $\mathcal{O}/\mathfrak{P}$.

Definition 7.5. The fixed field K^Z of the decomposition group $Z = Z_{\mathfrak{P}/\mathfrak{p}}$ is called the *decomposition field* of $\mathfrak{P}/\mathfrak{p}$. The fixed field K^T of T is the *inertia field* of $\mathfrak{P}/\mathfrak{p}$.

Since T is a subgroup of Z, K^Z is a subfield of K^T. Also K/k is Galois, hence

$$\mathfrak{p}\mathcal{O} = (\mathfrak{P}_1 \cdots \mathfrak{P}_g)^e \tag{7.3}$$

with $f_1 = \cdots = f_g = f$, say. So we have $[K : k] = n$,

$$n = efg. \tag{7.4}$$

A *Hasse diagram* is a diagram in which the inclusion between field extensions and their subsets is indicated by putting a subset at a lower level of the page and connecting them by a line and the line joining a field extension is labeled with the extension degree. For example, the following diagram describes a central theorem of Hilbert Theory.

Theorem 7.6. *If $\mathfrak{P}$ is a prime of K dividing a prime $\mathfrak{p}$ of k, we have the following Hasse diagram.*

$$
\begin{array}{ccccc}
K & & \mathcal{O} & & \\
e\Big| & & \Big| & & \mathfrak{P} \\
K^T & & \mathcal{O}^T & & \Big| \\
f\Big| & & \Big| & & \mathfrak{P}^T \\
K^Z & & \mathcal{O}^Z & & \Big| \\
g\Big| & & \Big| & & \mathfrak{P}^Z \\
k & & \mathfrak{o} & & \Big| \\
& & & & \mathfrak{p}
\end{array}
$$

Further,

1. $[K : K^T] = e$, $[K^T : K^Z] = f$, $[K^Z : k] = g$,

2. (a) $e(\mathfrak{P}^Z/\mathfrak{p}) = f(\mathfrak{P}^Z/\mathfrak{p}) = 1$,

    (b) $e(\mathfrak{P}^T/\mathfrak{P}^Z) = 1$, $f(\mathfrak{P}^T/\mathfrak{P}^Z) = f(\mathfrak{P}/\mathfrak{p}) = f$,

    (c) $e(\mathfrak{P}/\mathfrak{P}^T) = e(\mathfrak{P}/\mathfrak{p}) = e$, $f(\mathfrak{P}/\mathfrak{P}^T) = 1$.

Proof. Without loss of generality, we may assume that $\mathfrak{P} = \mathfrak{P}_1$.

We first prove that $[K^Z : k] = g$. By Galois theory, the degree of the extension $[K^Z : k] = [G : Z]$, the index of Z in G. Hence it is enough to show that there is a bijection between the g primes $\mathfrak{P}_1, \ldots, \mathfrak{P}_g$ and the cosets of G/Z. Given σ in G, let $\sigma(\mathfrak{P}) = \mathfrak{P}_j$. By definition, $\tau(\mathfrak{P}) = \mathfrak{P}$, for all τ in Z. Hence $\mu(\mathfrak{P}) = \mathfrak{P}_j$ for each μ in σZ, and thus we have a map from σZ to j or $\mathfrak{P}_j$. This map is surjective, because given $\mathfrak{P}_j$, there is a σ in G such that $\sigma(\mathfrak{P}) = \mathfrak{P}_j$. Moreover, $\sigma_1 \mathfrak{P} = \sigma_2 \mathfrak{P}$ implies that $\sigma_1^{-1} \sigma_2(\mathfrak{P}) = \mathfrak{P}$, i.e. $\sigma_1^{-1} \sigma_2 \in Z$, which is so if and only if $\sigma_1 Z = \sigma_2 Z$. This gives the required bijection.

The fact that $[K^Z : k] = g$, just proved, shows that the extension K/K^Z is normal of degree ef with Galois group Z. By definition, Z fixes $\mathfrak{P} = \mathfrak{P}_1$. Hence $\mathfrak{P}$ is the only prime in K dividing $\mathfrak{P}^Z$. But $e' = e(\mathfrak{P}/\mathfrak{P}^Z) \leq e(\mathfrak{P}/\mathfrak{p}) = e$ and $f' = f(\mathfrak{P}/\mathfrak{P}^Z) \leq f(\mathfrak{P}/\mathfrak{p}) = f$. Now by (7.3), $e' \cdot f' \cdot 1 = [K : K^Z] = ef$. This can happen only if $e = e'$, $f = f'$ and this also forces $e(\mathfrak{P}^Z/\mathfrak{p}) = 1$ and $f(\mathfrak{P}^Z/\mathfrak{p}) = 1$.

Now we show that $f(\mathfrak{P}/\mathfrak{P}^T) = 1$. Let $\overline{K} = \mathcal{O}/\mathfrak{P}$ and $\overline{K^T} = \mathcal{O}^T/\mathfrak{P}^T$. By definition, $f(\mathfrak{P}/\mathfrak{P}^T) = 1 \Leftrightarrow [\overline{K} : \overline{K^T}] = 1 \Leftrightarrow \mathrm{Gal}(\overline{K}/\overline{K^T}) = \{1\}$.

If for α in $\mathcal{O}$, $\overline{\alpha}$ denotes $\alpha \bmod \mathfrak{P}$, then $[\overline{K} : \overline{K^T}] = 1 \Leftrightarrow \forall \ \overline{\alpha}$ in $\mathcal{O}/\mathfrak{P}$, the polynomial $(x - \overline{\alpha})^m$ is in $(\mathcal{O}^T/\mathfrak{P}^T)[x]$ for some $m \geq 1$, because then each σ in $\mathrm{Gal}(\overline{K}/\overline{K^T})$ sends $\overline{\alpha}$ to a root of $(x - \overline{\alpha})^m$, which can only be $\overline{\alpha}$. So let $\alpha \in \mathcal{O}$.

The polynomial

$$\phi(x) = \prod_{\sigma \in T} (x - \sigma(\alpha))$$

is in $\mathcal{O}^T[x]$. If $\overline{\phi}(x)$ denotes the polynomial over $\overline{K}$ obtained from $\phi(x)$ by reducing its coefficients $\bmod \mathfrak{P}$, then $\overline{\phi}(x) \in \overline{K^T}[x]$. By definition of T, $\forall \sigma \in T$, $\sigma(\alpha) \equiv \alpha \pmod{\mathfrak{P}}$, i.e. $\overline{\sigma(\alpha)} = \overline{\alpha}$, hence $\overline{\phi}(x) = (x - \overline{\alpha})^m$ with $m = |T|$, proving that $f(\mathfrak{P}/\mathfrak{P}^T) = 1$. Since $f = f(\mathfrak{P}/\mathfrak{p}) = f(\mathfrak{P}/\mathfrak{P}^T)f(\mathfrak{P}^T/\mathfrak{P}^Z)f(\mathfrak{P}^Z/\mathfrak{p})$, it now follows that $f(\mathfrak{P}^T/\mathfrak{P}^Z) = f$, which implies that $[K^T : K^Z] \geq f$. But by equation (7.4), $[K^T : K^Z] = |Z/T|$ is a factor of f. Hence $[K^T : K^Z] = f$ which implies that $[K : K^T] = e$. The rest of the statements now follow easily from what we have already proved. $\square$

Corollary 7.7. *Let* $\overline{K} = \mathcal{O}/\mathfrak{P}$, $\overline{k} = \mathfrak{o}/\mathfrak{p}$. *Then* $\operatorname{Gal}(\overline{K}/\overline{k}) \cong Z/T$.

EXERCISE

Give counter-examples to show that Z and T need not be normal in $G = \operatorname{Gal}(K/k)$.

Corollary 7.8. *If* $Z = Z_{\mathfrak{P}/\mathfrak{p}}$ *is normal in* $G = \operatorname{Gal}(K/k)$, *then* $\mathfrak{p}$ *splits completely in* K^Z. *Further, if* T *is also normal in* G, *then each of the g prime divisors of* $\mathfrak{p}$ *in* K^Z *stays inert in* K^T, *and finally becomes the e-th power of a prime in* K.

Proof. If Z is normal in G, then K^Z/k is a normal extension. Hence for all primes in K^Z dividing $\mathfrak{p}$, the ramification indices and the residue class field degrees are equal to one, and therefore, there must be g primes in K^Z dividing $\mathfrak{p}$. Again, by hypothesis K^T/K^Z is a normal extension, so all the residue class field degrees are equal to f, hence the ramification indices are all equal to one. The last statement is now obvious. $\qquad\square$

Theorem 7.9. *(1)* K^Z *is the unique field* L, *intermediate between* k *and* K, *such that if* $\mathfrak{P}_L = \mathfrak{P} \cap L$, *then* $e(\mathfrak{P}_L/\mathfrak{p}) = f(\mathfrak{P}_L/\mathfrak{p}) = 1$, *and* $\mathfrak{P}$ *is the only prime in* K *dividing* $\mathfrak{P}_L$.

(2) K^T *is the unique field* L, *intermediate between* k *and* K, *such that for* $\mathfrak{P}_L = \mathfrak{P} \cap L$, $e(\mathfrak{P}_L/\mathfrak{p}) = 1$ *and* $\mathfrak{P}$ *is totally ramified over* $\mathfrak{P}_L$.

Proof. (1) Suppose L is a field with this property. Let $H = \{\sigma \in \operatorname{Gal}(K/k) \mid \sigma_{|L} = 1_L\}$. By Galois theory, $L = K^H$. If $\sigma \in H$, then $\sigma\mathfrak{P}_L = \mathfrak{P}_L$. Now $\mathfrak{P}$ being the only prime in K dividing $\mathfrak{P}_L$, it follows that $\sigma\mathfrak{P} = \mathfrak{P} \Rightarrow \sigma \in Z \Rightarrow H \subseteq Z \Rightarrow K^Z \subseteq L$. If $e(\mathfrak{P}_L/\mathfrak{p}) = f(\mathfrak{P}_L/\mathfrak{p}) = 1$, then $e(\mathfrak{P}/\mathfrak{P}_L) = e$ and $f(\mathfrak{P}/\mathfrak{P}_L) = f \Rightarrow [K : L] \geq ef \Rightarrow [L : k] \leq g$. This together with $[K^Z : k] = g$ and $K^Z \subseteq L$ implies that $L = K^Z$.

We leave the proof of this part as an exercise. $\qquad\square$

7.2   Higher Ramification Groups

Again we fix a normal extension K/k of number fields, a prime $\mathfrak{p}$ in k and a prime $\mathfrak{P}$ in K dividing $\mathfrak{p}$. For each non-negative integer j, we define the jth *ramification group* V_j of $\mathfrak{P}/\mathfrak{p}$ by

$$V_j = \{\sigma \in \operatorname{Gal}(K/k) \mid \sigma(\alpha) \equiv \alpha \bmod \mathfrak{P}^{j+1}, \forall \, \alpha \in \mathcal{O}\}.$$

The letter V is from the German word Verweigung for ramification.

Thus $V_0 = T$ and

$$V_0 \supseteq V_1 \supseteq V_2 \supseteq \cdots .$$

By the unique factorization theorem of Dedekind, if $\alpha \in \mathcal{O}$, $\alpha \neq 0$, then for some $j \geq 0$, $\alpha \in \mathfrak{P}^j$, but $\alpha \notin \mathfrak{P}^{j+1}$. Hence,

$$\bigcap_{j=1}^{\infty} \mathfrak{P}^j = \{0\}$$

and $\sigma(\alpha) \equiv \alpha \pmod{\mathfrak{P}^j}$ can hold for all $j \geq 0$ if and only if $\sigma = id$. This shows that

$$\bigcap_{j=0}^{\infty} V_j = \{1\}.$$

Since G is finite, $V_m = \{1\}$ for some $m \geq 0$. It is easy to see that each V_j is normal in Z, because if $\sigma \in Z$, $\tau \in V_j$, then $\forall\ \alpha$ in $\mathcal{O}$,

$$\tau(\sigma(\alpha)) - \sigma(\alpha) \in \mathfrak{P}^{j+1},$$

which implies that $(\sigma^{-1}\tau\sigma)(\alpha) - \alpha$ is in $\mathfrak{P}^{j+1}$. As a consequence, each V_j is normal in V_{j-1}.

Theorem 7.10. *(1) T/V_1 is isomorphic to a subgroup of the multiplicative group $\overline{K}^{\times}$, where $\overline{K} = \mathcal{O}/\mathfrak{P}$ is the residue field. Hence T/V_1 is cyclic and its order divides $q^f - 1$.*

(2) $\forall\ j \geq 2$, V_{j-1}/V_j is isomorphic to a subgroup of the additive group $\overline{K}$. Hence, its order is a prime power p^r $(r \geq 1)$, where $p = char\ \overline{K}$.

Proof. (1) Let $S = \mathfrak{o} \setminus \mathfrak{p}$. If $\mathfrak{o}' = S^{-1}\mathfrak{o}$, $\mathfrak{p}' = S^{-1}\mathfrak{p}$, $\mathfrak{P}' = S^{-1}\mathfrak{P}$, $\mathcal{O}' = S^{-1}\mathcal{O}$ and $\sigma \in G$. Then $\sigma(\alpha) \equiv \alpha \pmod{\mathfrak{P}^j}$, $\forall\ \alpha$ in $\mathcal{O}$ if and only if $\sigma(\alpha') \equiv \alpha'$ $(\bmod\ \mathfrak{P}^j)$, $\forall\ \alpha' \in \mathcal{O}'$. Hence, if

$$V_j' = \{\sigma \in G \mid \sigma(\alpha') \equiv \alpha' \pmod{\mathfrak{P}'^j}, \forall\ \alpha' \in \mathcal{O}'\},$$

then $V_j = V_j'$. Thus localizing at $\mathfrak{p}$, we may assume that $\mathfrak{P}$ is principal. Choose a uniformizing parameter at $\mathfrak{P}$, i.e. an element π in $\mathfrak{P} \setminus \mathfrak{P}^2$. Then $\mathfrak{P} = (\pi)$.

If $\sigma \in Z$, then $\sigma\mathfrak{P} = \mathfrak{P}$, so $\sigma(\pi) = a_\sigma \pi$ with a_σ in $\mathcal{O}$. The unique a_σ is not in $\mathfrak{P}$, otherwise we would have $\sigma\mathfrak{P} = \mathfrak{P}^2$. Thus we have a map

$$Z \ni \sigma \to \overline{a}_\sigma \in (\mathcal{O}/\mathfrak{P})^{\times}, \tag{7.5}$$

where $\overline{a}_\sigma$ is the coset of a_σ in $\mathcal{O}/\mathfrak{P}$. We claim that the restriction of this map to T is an isomorphism with kernel V_1. To prove this, let $\sigma, \tau \in T = V_0$. Then

$$a_{\sigma\tau}\pi = \sigma(\tau(\pi)) = \sigma(a_\tau \pi) = \sigma(a_\tau)\sigma(\pi) = \sigma(a_\tau)a_\sigma\pi,$$

which gives

$$a_{\sigma\tau} = \sigma(a_\tau)a_\sigma.$$

Because $\sigma \in T$, $\sigma(a_\tau) \equiv a_\tau \pmod{\mathfrak{P}}$. Hence,

$$a_{\sigma\tau} \equiv a_\tau a_\sigma \pmod{\mathfrak{P}}.$$

Thus the restriction of the map in (7.5) to T is a homomorphism. $\qquad\square$

To prove that the kernel is V_1, we shall need the following characterization of the subgroups V_j.

Lemma 7.11. *The groups $V_m(m \geq 1)$ can be characterized as follows:*

$$V_m = \{\sigma \in G \,|\, \sigma(\pi) \equiv \pi \pmod{\mathfrak{P}^{m+1}}\}.$$

Proof. We have only to show that

$$\sigma(\pi) \equiv \pi \pmod{\mathfrak{P}^{m+1}} \tag{7.6}$$

implies $\sigma(\alpha) \equiv \alpha \pmod{\mathfrak{P}^{m+1}}$, $\forall\, \alpha \in \mathcal{O}$. By Theorem 7.6, $\mathcal{O}^T/\mathfrak{P}^T \cong \mathcal{O}/\mathfrak{P}$. Hence, we may choose coset representatives of $\mathcal{O}/\mathfrak{P}$ from $\mathcal{O}^T$. Also the multiplication by π^m gives an isomorphism of the additive groups

$$\mathcal{O}/\mathfrak{P} \cong \mathfrak{P}^m/\mathfrak{P}^{m+1}.$$

Hence, mod $\mathfrak{P}^{m+1}$, each element α of $\mathcal{O}$ has a representation

$$\alpha = a_0 + a_1\pi + \cdots + a_m\pi^m \;(a_j \in \mathcal{O}^T).$$

If (7.6) holds, then for σ in $V_0 = T$,

$$\begin{aligned}
\sigma(\alpha) &= \sigma(a_0) + \sigma(a_1)\sigma(\pi) + \cdots + \sigma(a_m)\sigma(\pi^m) \\
&= a_0 + a_1\pi + \cdots + a_m\pi^m \\
&\equiv \alpha \pmod{\mathfrak{P}^{m+1}}.
\end{aligned}$$

This proves the lemma. $\qquad\square$

To show that the kernel of the map $\sigma \to \bar{a}_\sigma$ is V_1, let $\sigma \in V_0$. By Lemma 7.11, $\sigma \in V_1 \Leftrightarrow \sigma(\pi) \equiv \pi \pmod{\mathfrak{P}^2} \Leftrightarrow a_\sigma\pi \equiv \pi \pmod{\mathfrak{P}^2} \Leftrightarrow a_\sigma \equiv 1 \pmod{\mathfrak{P}} \Leftrightarrow \sigma$ is in the kernel.

(2) Now let $j \geq 2$. If $\sigma \in V_{j-1}$, then $\sigma(\pi) \equiv \pi \pmod{\mathfrak{P}^j}$. Hence

$$\sigma(\pi) - \pi = b_\sigma\pi^j, \tag{7.7}$$

for a unique b_σ in $\mathcal{O}$.

If τ is another element of V_{j-1}, then

$$\tau(\pi) - \pi = b_\tau \pi^j \ (b_\tau \in \mathcal{O}), \tag{7.8}$$

and by (7.7) and (7.8), we have

$$\begin{aligned}
b_{\sigma\tau}\pi^j &= \sigma\tau(\pi) - \pi \\
&= \sigma(b_\tau \pi^j + \pi) - \pi \\
&= \sigma(b_\tau)(\sigma(\pi))^j + (\sigma(\pi) - \pi) \\
&= \sigma(b_\tau)(\pi + b_\sigma \pi^j)^j + b_\sigma \pi^j.
\end{aligned}$$

Dividing throughout by π^j, we get

$$b_{\sigma\tau} = \sigma(b_\tau)(1 + b_\sigma \pi^{j-1})^j + b_\sigma. \tag{7.9}$$

Because $j-1 \geq 1$, $\sigma(b_\tau) \equiv b_\tau \pmod{\mathfrak{P}}$. Hence reducing equation (7.9) mod $\mathfrak{P}$, we get

$$\bar{b}_{\sigma\tau} = \bar{b}_\sigma + \bar{b}_\tau.$$

This gives a homomorphism from V_{j-1} to $\mathcal{O}/\mathfrak{P}$.

To show that the kernel of this map is V_j, let $\sigma \in V_{j-1}$. By Lemma 7.11, $\sigma \in V_j \Leftrightarrow \sigma(\pi) \equiv \pi \pmod{\mathfrak{P}^{j+1}} \Leftrightarrow \sigma(\pi) - \pi = b_\sigma \pi^j \in \mathfrak{P}^{j+1} \Leftrightarrow b_\sigma \equiv 0 \pmod{\mathfrak{P}} \Leftrightarrow \sigma$ belongs to the kernel.

The order of the additive group $\mathcal{O}/\mathfrak{P}$ is $N_{L/\mathbb{Q}}(\mathfrak{P}) = p^s$ for some $s \geq 1$. Hence $|V_{j-1}/V_j| = p^r$ with $0 \leq r \leq s$.

Theorem 7.12. *The map*

$$Z \ni \sigma \to \bar{\sigma} \in \mathrm{Gal}(\overline{K}/\bar{k})$$

is a group homomorphism with kernel T. Hence, T is normal in Z, and the quotient group $Z/T \cong \mathrm{Gal}(\overline{K}/\bar{k})$.

Proof. It is obvious that $\sigma \to \bar{\sigma}$ is a group homomorphism for $\overline{\sigma\tau}(\alpha + \mathfrak{P}) = \sigma\tau(\alpha) + \mathfrak{P} = \bar{\sigma} \circ \bar{\tau}(\alpha + \mathfrak{P})$. By definition, its kernel is

$$\{\sigma \in Z \mid \sigma(\alpha) \equiv \alpha \pmod{\mathfrak{P}}, \forall\, \alpha \in \mathcal{O}\},$$

which is T. Thus the map $\sigma \to \bar{\sigma}$ induces an injective homomorphism $Z/T \to \mathrm{Gal}(\overline{K}/\bar{k})$. But Z/T and $\mathrm{Gal}(\overline{K}/\bar{k})$ both have f elements. Hence $Z/T \cong \mathrm{Gal}(\overline{K}/\bar{k})$. $\qquad\square$

7.3 The Frobenius Map

The multiplicative group $\overline{k}^{\times}$ of all the nonzero elements of the residue field $\overline{k} = \mathfrak{o}/\mathfrak{p}$ is a cyclic group of order $q - 1$, where q is the cardinality of $\overline{k}$. Hence for each a in $\overline{k}$, $a^q = a$. Moreover, since $(\alpha + \beta)^q = \alpha^q + \beta^q$ for all α, β in $\overline{K}$, the map $\Phi : \overline{K} \to \overline{K}$, given by

$$\Phi(x) = x^q$$

is a $\overline{k}$-linear map. Thus, $\Phi \in \mathrm{Gal}(\overline{K}/\overline{k})$. If α is a generator of the cyclic group $\overline{K}^{\times}$, which is of order $q^f - 1$, then $\alpha, \Phi(\alpha), \Phi^2(\alpha), \ldots, \Phi^{f-1}(\alpha)$ are all distinct, because otherwise, $\Phi^j(\alpha) = \alpha$ for some j $(0 < j < f)$. That means that $\alpha^{q^j} = \alpha \Rightarrow \alpha^{q^j - 1} = 1$, i.e. in the group $\overline{K}^{\times}$, $\mathrm{ord}(\alpha) = q^f - 1 \leq q^j - 1 \Rightarrow j \geq f$, a contradiction. Since $\mathrm{Gal}(\overline{K}/\overline{k}) = [\overline{K} : \overline{k}] = f$, the distinct powers, $1, \Phi, \Phi^2, \ldots, \Phi^{f-1}$ are all the elements of $\mathrm{Gal}(\overline{K}/\overline{k})$. Hence $\mathrm{Gal}(\overline{K}/\overline{k})$ is a cyclic group generated by Φ.

Definition 7.13. The generator Φ of the cyclic group $\mathrm{Gal}(\overline{K}/\overline{k})$, given by

$$\Phi(x) = x^q$$

is called the *Frobenius map* over $\overline{k}$.

Clearly Φ depends only on the ground field $\overline{k}$ of the extension $\overline{K}/\overline{k}$.

Definition 7.14. A Galois extension K/k is a *cyclic extension*, an *Abelian extension* or a *solvable extension* according as the Galois group $\mathrm{Gal}(K/k)$ is cyclic, Abelian or solvable.

Recall that a group G is *solvable* if there is a chain of subgroups

$$G = G_0 \supseteq G_1 \supseteq \cdots \supseteq G_m = \{1\}$$

such that G_{j+1} is normal in G_j and the quotient group G_j/G_{j+1} is Abelian. In particular, every Abelian group is solvable.

Theorem 7.15. *The decomposition group $Z = Z_{\mathfrak{P}/\mathfrak{p}}$ is solvable.*

Proof. Consider the chain of subgroup

$$Z \supseteq T = V_0 \supseteq V_1 \supseteq \cdots \supseteq V_m = \{1\}.$$

By Theorems 7.10 and 7.12, the quotients Z/T and V_j/V_{j+1} $(j \geq 0)$ are all Abelian groups. $\qquad \square$

We have seen that the order of the quotient group V_0/V_1 divides $q^f - 1 = (q - 1)(q^{f-1} + \cdots + q + 1)$. More is true if K/k is Abelian.

Theorem 7.16. *If K/k is Abelian, then $|V_0/V_1|$ divides $q - 1$.*

Proof. Choose τ in $T = V_0$ such that its coset in V_0/V_1 generates the cyclic group V_0/V_1, and choose σ in Z such that its coset in $Z/T \cong \text{Gal}(\overline{K}/\overline{k})$ is mapped to the Frobenius map $\Phi : \overline{K} \to \overline{K}$, given by $\Phi(x) = x^q$.

Let π be a uniformizing parameter at $\mathfrak{P}$ and $\sigma(\pi) = a_\sigma \pi$, $\tau(\pi) = a_\tau \pi$. To simplify notation, let us put $a_\sigma = a$ and $a_\tau = b$. Clearly, $a \notin \mathfrak{P}$, otherwise $\sigma(\mathfrak{P}) = \mathfrak{P} \subseteq \mathfrak{P}^2$, which is not true. By our hypothesis, G is Abelian. Hence $b\pi = \tau(\pi) \Rightarrow \sigma(b)\sigma(\pi) = \sigma\tau(\pi) = \tau\sigma(\pi) = \tau(a\pi) = \tau(a)\tau(\pi)$, which gives $\sigma(b)a\pi = \tau(a)b\pi$, i.e.

$$a\sigma(b) = b\tau(a). \tag{7.10}$$

By our choice of σ and τ, this gives

$$b^q a \equiv ab \pmod{\mathfrak{P}}.$$

Since $a, b \notin \mathfrak{P}$, this gives

$$b^{q-1} \equiv 1 \pmod{\mathfrak{P}}.$$

Hence, $\text{ord}(\tau) = \text{ord}(\overline{b})$ in $(\mathcal{O}/\mathfrak{P})^\times$ is a factor of $q - 1$. $\qquad\square$

We shall need two results, one of them from Galois theory. We shall only state them, leaving their proofs as exercises. Recall that the *composition field* $K_1 K_2$ of K_1 and K_2 is the smallest field containing K_1 and K_2.

Proposition 7.17. *If K_1/k, K_2/k are Galois extensions with Galois groups G_1, G_2 and $K = K_1 \cap K_2$, then $K_1 K_2/K$ is Galois. The Galois group $\text{Gal}(K_1 K_2/K)$ is isomorphic to the subgroup*

$$\{(\sigma_1, \sigma_2) \in G_1 \times G_2 \mid \sigma_{1|K} = \sigma_{2|K}\}$$

of $G_1 \times G_2$. Thus if $K_1 \cap K_2 = k$, then $\text{Gal}(K_1 K_2/K) \cong G_1 \times G_2$. Moreover, if K_1/k and K_2/k are Abelian, then so is $K_1 K_2/K$.

Proposition 7.18. *Suppose K_1, K_2 are Galois extensions of k and $K = K_1 K_2$. A prime $\mathfrak{p}$ of k ramifies in K if and only if $\mathfrak{p}$ ramifies in K_1 or K_2.*

[*Hint*: Let $\mathfrak{P}$ be a prime of K dividing $\mathfrak{p}$. Let $\mathfrak{P}_j = \mathfrak{P} \cap K_j$. Inject the group $T_{\mathfrak{P}/\mathfrak{p}}$ into $T_{\mathfrak{P}_1/\mathfrak{p}} \times T_{\mathfrak{P}_2/\mathfrak{p}}$.]

We also need the following fact from group theory, whose proof is left as an exercise.

Proposition 7.19. *Suppose G is an Abelian group of order a prime power p^m ($m \geq 1$). Then G is cyclic if and only if G has a unique subgroup of order p^{m-1}.*

7.4 Ramification in Cyclic Extensions

A cyclic extension is Abelian but not conversely. An example of an Abelian extension, which is not cyclic is the following. Let $K = \mathbb{Q}(\sqrt{2}, \sqrt{3})$. Then $K/\mathbb{Q}$ is Galois with Galois group

$$\text{Gal}(K/\mathbb{Q}) \cong \mathbb{Z}/2\mathbb{Z} \times \mathbb{Z}/2\mathbb{Z},$$

which is Abelian but not cyclic.

In this section, all extensions are assumed to be Abelian. Thus $Z_{\mathfrak{P}/\mathfrak{p}}$, $T_{\mathfrak{P}/\mathfrak{p}}$, $e(\mathfrak{P}/\mathfrak{p})$, $f(\mathfrak{P}/\mathfrak{p})$, etc. depend only on $\mathfrak{p}$, not on $\mathfrak{P}$. For example, if $\mathfrak{P}$, $\mathfrak{Q}$ are two primes in K dividing $\mathfrak{p}$, then $\mathfrak{Q} = \tau\mathfrak{P}$ for τ in G. Therefore $\sigma\mathfrak{P} = \mathfrak{P} \Leftrightarrow \tau\sigma\mathfrak{P} = \tau\mathfrak{P} \Leftrightarrow \sigma\tau\mathfrak{P} = \tau\mathfrak{P} \Leftrightarrow \sigma\mathfrak{Q} = \mathfrak{Q}$, hence $Z_{\mathfrak{P}/\mathfrak{p}} = Z_{\mathfrak{Q}/\mathfrak{p}}$. Thus, we may denote these objects simply by $Z_{\mathfrak{p}}$, $T_{\mathfrak{p}}$, $e_{\mathfrak{p}}$, $f_{\mathfrak{p}}$, etc.

We now study the ramification of a rational prime p in an Abelian extension $K/\mathbb{Q}$ of degree p^m ($m \geq 1$). It is therefore given that Z_p, T_p and the higher ramification groups

$$V_j = \{\sigma \in \text{Gal}(K/\mathbb{Q}) \,|\, \sigma(\alpha) \equiv \alpha \bmod \mathfrak{P}^{j+1}\},$$

where $\mathfrak{P}$ is any prime in K dividing p, depend only on p. Since $G = \text{Gal}(K/\mathbb{Q})$ is of order p^m, all the subgroups of G and their quotients have order p^r ($r \leq m$). We have seen that the quotient V_0/V_1 is isomorphic to the multiplicative group $\overline{K}^{\times}$, where $\overline{K}$ is the residue field $\mathcal{O}/\mathfrak{P}$ of cardinality p^r for some $r \geq 1$. Since $|V_0/V_1|$ is also a power of p, the order $|V_0/V_1|$ can divide $p^r - 1 = |\overline{K}^{\times}|$ (Theorem 7.10) only if it is one. Hence, we have proved the following result.

Theorem 7.20. *Suppose $K/\mathbb{Q}$ is Abelian of degree p^m ($m \geq 1$). If V_j ($j \geq 0$) are the ramification groups for the prime p of $\mathbb{Q}$, then $V_0 = V_1$.*

Recall that p ramifies in a finite extension L of $\mathbb{Q}$ if and only if $p|d_L$, the discriminant of L. On the other hand, by Minkowski's Theorem on the discriminant, $d_L = 1$ if and only if $L = \mathbb{Q}$. Hence, in every proper extension of $\mathbb{Q}$, at least one prime must ramify.

If $\mathfrak{P}$ is a prime of a Galois extension K of $\mathbb{Q}$ dividing p and $T = T_{\mathfrak{P}/\mathfrak{p}}$, then no prime over p ramifies in the fixed field K^T of T. Thus if p is the only prime to ramify in K, then by Theorem 7.6, $K^T = \mathbb{Q}$.

By Theorem 7.20, we at once obtain the following.

Theorem 7.21. *If p is the only prime to ramify in a Galois extension $K/\mathbb{Q}$, then p is totally ramified in K. Moreover, $f(\mathfrak{P}/\mathfrak{p}) = 1$ and $\mathcal{O}/\mathfrak{P} \cong \mathbb{F}_p$, the field of p elements.*

For $j \geq 2$, V_{j-1}/V_j is isomorphic to the additive group of the residue field $\overline{K} = \mathcal{O}/\mathfrak{P} \cong \mathbb{F}_p$. Hence the following is obvious.

Theorem 7.22. *If p is the only prime to ramify in an Abelian extension $K/\mathbb{Q}$ of degree p^m $(m \geq 1)$, then for $j \geq 2$, the order $|V_{j-1}/V_j|$ is 1 or p.*

Theorem 7.23. *Suppose $p > 2$ is the only prime to ramify in an Abelian extension K of $\mathbb{Q}$ of degree p. Then the ramification group V_2 (of p) is $\{1\}$.*

Proof. Let $\mathfrak{P}$ be a prime of K dividing p. By localizing, we may assume that $\mathfrak{P} = (\pi)$ is a principal ideal. By Theorem 7.21, p is totally ramified in K, i.e. $(p) = (\pi^p)$. Since $\pi \in \mathcal{O}_K$ and $\deg_{\mathbb{Q}}(\pi) > 1$ and $[K : \mathbb{Q}] = p$, π is a root of a monic polynomial

$$f(x) = a_0 + a_1 x + \cdots + x^p$$

over $\mathbb{Z}$.

Recall the *discrete valuation*

$$v = v_{\mathfrak{P}} : K^\times \to \mathbb{Z}$$

at the prime $\mathfrak{P}$, namely that $v_{\mathfrak{P}}(\alpha)$ is the exponent (positive, negative or zero) of $\mathfrak{P}$ in the unique factorization of the principal (fractional) ideal (α) into products of powers of distinct primes. The map v has the following properties.

$$v(\alpha\beta) = v(\alpha) + v(\beta), \text{ and} \tag{7.11}$$
$$\text{if } v(\alpha) \neq v(\beta), \text{ then } v(\alpha + \beta) = \min(v(\alpha), v(\beta)).$$

Of course, (7.11) extends for $v(\alpha_1 \ldots \alpha_n)$ and $v(\alpha_1 + \cdots + \alpha_n)$ in an obvious way.

The Galois group $\mathrm{Gal}(K/\mathbb{Q})$ has order p, so it has no nontrivial subgroup. Since p is totally ramified, $T = G$ and by Theorem 7.20, $T = V_0 = V_1$. Let $j \geq 2$ be the smallest integer such that $V_{j-1} = G$ and $V_j = \{1\}$. We shall show that $j = 2$.

First note that

$$v(f'(\pi)) = j(p-1). \tag{7.12}$$

This is so because

$$f'(\pi) = \prod_{\sigma \in G, \sigma \neq id} (\pi - \sigma(\pi)) = \prod_{\sigma \in V_{j-1}\backslash V_j} (\pi - \sigma(\pi)). \tag{7.13}$$

Since

$$V_{j-1} = \{\sigma \in G \mid \sigma(\alpha) \equiv \alpha \pmod{\mathfrak{P}^j}, \forall\, \alpha \in \mathcal{O}\},$$

it is clear that $\forall\, \sigma$ in $V_{j-1} \setminus V_j$,

$$v(\pi - \sigma(\pi)) = j. \tag{7.14}$$

Therefore, (7.12) follows at once from (7.11), (7.13) and (7.14).

Setting $\alpha_j = ja_j\pi^{j-1}$ we also have

$$f'(\pi) = a_1 + 2a_2\pi + \cdots + (p-1)a_{p-1}\pi^{p-2} + p\pi^{p-1}$$
$$= \alpha_1 + \alpha_2 + \cdots + \alpha_p,$$

say. If we write a nonzero coefficient $a = ja_j$ of $f'(\pi)$ as $a = bp^r$ with $(b,p) = 1$, then in view of p being totally ramified, i.e. $v(p) = p$, it follows that $p \,|\, v(a)$. Hence

$$v(\alpha_i) \equiv i - 1 \pmod{p},$$

in particular, $v_p(\alpha_i)$ are all unequal. Thus by (7.11), $v(f'(\pi)) = \min v(\alpha_i)$. This implies that

$$v(\alpha_p) = v(p\pi^{p-1}) = 2p - 1 \geq v(f'(\pi)) = j(p-1).$$

This can hold for $j \geq 2$ only if $j = 2$. [This conclusion is not true, if $p = 2$.] $\square$

Theorem 7.24. *Suppose $K/\mathbb{Q}$ is an Abelian extension of degree p^m $(m \geq 1)$. If $p > 2$, and p is the only prime to ramify in K, then $K/\mathbb{Q}$ is cyclic.*

Proof. By induction on m. If $m = 1$, there is nothing to prove. So suppose $m > 1$. By Proposition 7.19, it suffices to show that the only subgroup of $G = \mathrm{Gal}(K/\mathbb{Q})$ of order p^{m-1} is V_2.

Let H be a subgroup of order p^{m-1}. Let $K' = K^H$ be the fixed field of H. Since the index $[G : H] = p$, by Galois theory, $[K' : \mathbb{Q}] = p$ and the Galois group $G' = \mathrm{Gal}(K'/\mathbb{Q}) \cong G/H$. Let V_j' be the ramification groups for p relative to the extension $K'|\mathbb{Q}$. By Theorem 7.23, $V_2' = \{1\}$. Since V_j' is the image of V_j in G/H, this implies that $V_2 \subseteq H$. Since p is totally ramified, $G = T = V_0$. By Theorem 7.20, $V_1 = V_0 = G$. By Theorem 7.22, $|V_1/V_2| = 1$ or p. Since $V_2 \subseteq H \not\subseteq G = V_1$, this shows that $H = V_2$. $\square$

7.5   The Artin Symbol

Let K/k be a Galois extension (not necessarily Abelian) of number fields with Galois group G. Let $\mathfrak{P}$ be a prime of K, dividing a fixed prime $\mathfrak{p}$ of k. Recall the definition of the decomposition group $Z = Z_{\mathfrak{P}/\mathfrak{p}}$:

$$Z = \{\sigma \in G \,|\, \sigma\mathfrak{P} = \mathfrak{P}\}.$$

If $\mathfrak{P}_1$, $\mathfrak{P}_2$ are two prime divisors of $\mathfrak{p}$ in K, then $\mathfrak{P}_2 = \tau\mathfrak{P}_1$ for some τ in G. For $j = 1, 2$, let $Z_j = Z_{\mathfrak{P}_j/\mathfrak{p}}$. Then trivially,

$$Z_2 = \tau^{-1}Z_1\tau.$$

From now on, let $\mathfrak{p}$ be unramified ($\Leftrightarrow \mathfrak{p} \nmid \mathfrak{d}_{K/k}$) so that the inertia group $T_{\mathfrak{P}/\mathfrak{p}} = \{1\}$, $\forall \, \mathfrak{P}$ in $\mathcal{O}$ dividing $\mathfrak{p}$. Then the canonical generators $\sigma_{\mathfrak{P}}$ of the cyclic group $Z_{\mathfrak{P}/\mathfrak{p}}$ under the isomorphism

$$Z_{\mathfrak{P}/\mathfrak{p}} \cong \mathrm{Gal}((\mathcal{O}/\mathfrak{P})/(\mathfrak{o}/\mathfrak{p})) \tag{7.15}$$

for all $\mathfrak{P}/\mathfrak{p}$ are conjugates, and in fact, form a conjugacy class in G. We denote it by $\left(\frac{K/k}{\mathfrak{p}} \right)$.

Definition 7.25. The *Artin symbol* is the conjugacy class

$$\left(\frac{K/k}{\mathfrak{p}} \right)$$

of $G = \mathrm{Gal}(K/k)$.

Now assume that G is Abelian, so that $\left(\frac{K/k}{\mathfrak{p}} \right)$ consists of a single element σ of G. It is characterized by the property:

$$\sigma(\alpha) \equiv \alpha^{N_{k/\mathbb{Q}}(\mathfrak{p})} \pmod{\mathfrak{P}}, \forall \, \alpha \in \mathcal{O}_K,$$

where $\mathfrak{P}$ is any prime of K dividing $\mathfrak{p}$. In other words, for the Abelian extension K/k, the Artin symbol is the "pullback" under the isomorphism (7.15) of the Frobenius automorphism $\Phi_{\mathfrak{p}}$ of $\mathcal{O}/\mathfrak{P}$ over $\mathfrak{o}/\mathfrak{p}$ defined by

$$\Phi_{\mathfrak{p}}(x) = x^{N_{k/\mathbb{Q}}(\mathfrak{p})}$$

for x in $\mathcal{O}/\mathfrak{P}$.

In the Abelian case, recall our notation $Z_{\mathfrak{p}}$, $T_{\mathfrak{p}}$, $e_{\mathfrak{p}}$, $f_{\mathfrak{p}}$, etc. We then have

$$[K : k] = e_{\mathfrak{p}} f_{\mathfrak{p}} g_{\mathfrak{p}}.$$

Since, by our assumption, $e_{\mathfrak{p}} = 1$, $[K : k] = f_{\mathfrak{p}} g_{\mathfrak{p}}$. Therefore, $\mathfrak{p}$ splits completely in $K \Leftrightarrow f_{\mathfrak{p}} = 1 \Leftrightarrow \mathrm{Gal}((\mathcal{O}/\mathfrak{P})/(\mathfrak{o}/\mathfrak{p})) = \{1\} \Leftrightarrow$ the Artin symbol $\left(\frac{K/k}{\mathfrak{p}} \right) = 1$, the identity element of $\mathrm{Gal}(K/k)$. Therefore, we have proved the following result.

Theorem 7.26. *Suppose K/k is an Abelian extension of number fields and $\mathfrak{p}$ a prime of k, unramified in K. Then $\mathfrak{p}$ splits completely in K if and only if the Artin symbol $\left(\frac{K/k}{\mathfrak{p}} \right) = 1$.*

We now illustrate this theory with an application to the simplest nontrivial example.

7.6 Quadratic Fields

Let $d \neq 0, 1$ be a square-free integer and $K = \mathbb{Q}(\sqrt{d})$. The quadratic extension $K/\mathbb{Q}$ is Abelian with Galois group

$$G = \mathrm{Gal}(K/\mathbb{Q}) \cong \{\pm 1\}.$$

Let p be an odd prime with $(p, d) = 1$, so that p is unramified in K. The Artin symbol $\sigma = \left(\frac{K/k}{\mathfrak{p}} \right)$ is the element of G characterized by

$$\sigma(\alpha) \equiv \alpha^p \pmod{\mathfrak{P}}, \forall \, \alpha \in \mathcal{O}_K, \tag{7.16}$$

$\mathfrak{P}$ being any prime divisor of p in K. We know that

$$\mathcal{O}_K = \mathbb{Z} \oplus \mathbb{Z}\omega,$$

where

$$\omega = \begin{cases} \sqrt{d} & \text{if } d \equiv 2, 3 \pmod 4 \\ \frac{1+\sqrt{d}}{2} & \text{if } d \equiv 1 \pmod 4 \end{cases}$$

Since $\sigma_{|\mathbb{Q}} = id$, (7.16) is, therefore, equivalent to

$$\sigma(\omega) \equiv \omega^p \pmod{\mathfrak{P}}.$$

Now $2 \notin \mathfrak{P}$, hence it is a unit in $\mathcal{O}/\mathfrak{P}$. So (7.16) is equivalent to

$$\sigma(\sqrt{d}) \equiv (\sqrt{d})^p \pmod{\mathfrak{P}}.$$

Therefore, $\sigma = id \Leftrightarrow$

$$\sqrt{d} \equiv (\sqrt{d})^p \pmod{\mathfrak{P}}. \tag{7.17}$$

Because $(p, d) = 1$, $\sqrt{d} \notin \mathfrak{P}$, hence after cancelling $\sqrt{d}$, (7.17) is equivalent to

$$d^{\frac{p-1}{2}} \equiv 1 \pmod{\mathfrak{P}}.$$

Recall the *Legendre symbol*, which is a homomorphism

$$\chi : \mathbb{F}_p^\times \to \mathbb{F}_p^\times$$

given by

$$\chi(a) = \left(\frac{a}{p} \right) = a^{\frac{p-1}{2}} = \begin{cases} 1 & \text{if } a \in \mathbb{F}_p^{\times 2} \\ -1 & \text{otherwise} \end{cases}$$

This shows that $\left(\frac{K/\mathbb{Q}}{p} \right) = 1 \Leftrightarrow$ the Legendre symbol $\left(\frac{d}{p} \right) = 1$. But $\left(\frac{K/\mathbb{Q}}{p} \right)$ and $\left(\frac{d}{p} \right)$ can only be ± 1. Hence, we have proved the following fact.

Theorem 7.27. *For the quadratic extension $K = \mathbb{Q}(\sqrt{d})$, the Artin symbol $\left(\frac{K/\mathbb{Q}}{p}\right)$ is the same as the Legendre symbol $\left(\frac{d}{p}\right)$.*

If $p \mid d$, it is convenient to define the Legendre symbol $\left(\frac{d}{p}\right) = 0$. We have the following fact.

Corollary 7.28. *Suppose $d \neq 0, 1$ is a square-free integer and $K = \mathbb{Q}(\sqrt{d})$. For an odd prime p,*

1. *p is ramified in $K \Leftrightarrow \left(\frac{d}{p}\right) = 0$,*

2. *p splits in $K \Leftrightarrow \left(\frac{d}{p}\right) = 1$ and*

3. *p stays inert in $K \Leftrightarrow \left(\frac{d}{p}\right) = -1$.*

Remark 7.29. In the quadratic extension $K = \mathbb{Q}(\sqrt{d})$, p splits completely $\Leftrightarrow p$ splits.

The following was conjectured by Fermat in 1640 and proved by Euler in 1754.

Corollary 7.30. *An odd prime p is a sum of two squares if and only if $p \equiv 1$ (mod 4).*

Proof. The prime $p = x^2 + y^2 = (x + iy)(x - iy)$ if and only if p splits in $\mathbb{Q}(\sqrt{-1}) \Leftrightarrow \left(\frac{-1}{p}\right) = (-1)^{\frac{p-1}{2}} = 1 \Leftrightarrow p \equiv 1$ (mod 4). $\qquad\qquad \square$

7.7   The Artin Map

Let $\mathfrak{d} = \mathfrak{d}_{K/k}$ be the relative discriminant of an Abelian extension K/k of number fields. Fix an integral ideal $\mathfrak{m}$, divisible by all the prime divisors in k of $\mathfrak{d}$. Let $\mathfrak{I}(\mathfrak{m})$ be the multiplicative group of fractional ideals

$$\mathfrak{a} = \mathfrak{p}_1^{a_1} \cdots \mathfrak{p}_r^{a_r} \ (a_j \in \mathbb{Z}) \tag{7.18}$$

with all $(\mathfrak{p}_j, \mathfrak{m}) = 1$. We extend the Artin symbol $\left(\frac{K/k}{\mathfrak{p}}\right)$ to a group homomorphism,

$$\alpha_{K/k} : \mathfrak{I}(\mathfrak{m}) \to \mathrm{Gal}(K/k),$$

again called the *Artin map*, as follows. For $\mathfrak{a}$ as in (7.18), put

$$\alpha_{K/k}(\mathfrak{a}) = \left(\frac{K/k}{\mathfrak{a}} \right)$$

$$= \prod_{j=1}^{r} \left(\frac{K/k}{\mathfrak{p}_j} \right)^{a_j}.$$

Class field theory is the study of the Artin map $\alpha_{K/k}$. This map carries a great deal of information about the arithmetic of the relative extension K/k. For example, we may restate Theorem 7.26 as follows.

Theorem 7.31. *Let K/k be an Abelian extension of number fields with relative discriminant $\mathfrak{d}$. An unramified prime $\mathfrak{p}$ of k splits completely in $K \Leftrightarrow \mathfrak{p} \in \mathrm{Ker}(\alpha_{K/k})$, where $\alpha_{K/k}$ is the Artin map*

$$\alpha_{K/k} : \mathfrak{I}(\mathfrak{d}) \to \mathrm{Gal}(K/k).$$

We now ask the following question. Suppose K/k is an Abelian extension of number fields and $\mathfrak{p}$ is an unramified prime of k, so that $[K : k] = f_{\mathfrak{p}} g_{\mathfrak{p}}$. In particular, $g = g_{\mathfrak{p}}$ is a factor of $[K : k]$. Conversely, given a factor g of $[K : k]$, is there a prime $\mathfrak{p}$ in k, such that $g = g_{\mathfrak{p}}$? The answer is, in general no, because of the following fact.

Theorem 7.32. *Suppose K/k is an Abelian extension, such that there is a prime $\mathfrak{p}$ in k which stays prime in K, then K/k is cyclic.*

Proof. If $\mathfrak{p}$ stays prime in K, i.e. if $\mathfrak{p}$ is inert, then $f_{\mathfrak{p}} = [K : k]$. Therefore, $\mathrm{Gal}(\overline{K}/\overline{k})$ has order $f = [K : k]$. But there is a surjective map from $\mathrm{Gal}(K/k)$ to the cyclic group $\mathrm{Gal}(\overline{K}/\overline{k})$ of order $f = [K : k]$, which is also the order of $\mathrm{Gal}(K/k)$. Hence this map is also injective, and therefore, $\mathrm{Gal}(K/k)$ is isomorphic to the cyclic group $\mathrm{Gal}(\overline{K}/\overline{k})$. $\qquad\square$

Example 7.33. No prime stays prime in $K = \mathbb{Q}(\sqrt{2}, \sqrt{3})$, because $K/\mathbb{Q}$ is Abelian but not cyclic. Thus the answer to the above question with $g = 1$ is no.

EXERCISES

1. Prove Proposition 7.17.

2. Prove Proposition 7.18.

3. Prove Proposition 7.19.

8

Cyclotomic Fields

Beyond quadratic extensions, the cyclotomic extensions are the simplest and best understood ones. What makes them even more important is the fact that any Abelian extension is a subextension of a cyclotomic extension. In the next chapter, we shall give a proof of this celebrated theorem. It was stated by Kronecker in 1853 and proved by Weber in 1887. Actually, the first complete proof is due to Hilbert [24]. In 1911, after some unsuccessful attempts, Weber did eventually succeed in providing a correct proof of his own. Several other proofs of this important theorem have appeared in the literature, some only a few years ago.

8.1 Cyclotomic Fields

Fix an integer $m \geq 1$. The m-th *roots of unity* are the roots in $\mathbb{C}$, of the polynomial $x^m - 1$. They form a cyclic group μ_m of order m, generated by,

$$\zeta = \zeta_m = e^{2\pi i/m} = \cos\left(\frac{2\pi}{m}\right) + i\sin\left(\frac{2\pi}{m}\right). \tag{8.1}$$

The *primitive m-th roots of unity* are the generators of the cyclic group μ_m. Thus η is a primitive m-th root of unity if and only if

$$\eta = \zeta^a, \ (a, m) = 1.$$

Let $\phi(m)$ be the *Euler totient function*

$$\phi(m) = \#\{a \in \mathbb{N} \mid 1 \leq a \leq m, \ \gcd(a, m) = 1\}.$$

We have seen in Chapter 5, that

$$\phi(m) = m \prod_{p \mid m} \left(1 - \frac{1}{p}\right).$$

In particular,

$$\phi(p) = p - 1, \ \phi(p^r) = (p-1)p^{r-1} \text{ and } \phi(2^r) = 2^{r-1}.$$

In terms of the totient function, we may say that there are exactly $\phi(m)$ primitive m-th roots of unity. We shall exclude the trivial case $m = 1$ and 2, and unless stated to the contrary, assume that $m > 2$. The symbol ζ_m will stand for $\cos\left(\frac{2\pi}{m}\right) + i\sin\left(\frac{2\pi}{m}\right)$, whereas ζ will be any m-th root of unity.

Definition 8.1. The m-th *cyclotomic field* is the number field $K = \mathbb{Q}(\zeta_m)$.

The following statements are obvious.

1. If $n|m$, then $\mathbb{Q}(\zeta_n) \subseteq \mathbb{Q}(\zeta_m)$, because $\zeta_n = \zeta_m^{m/n}$.

2. Let $d = $ g.c.d. (m, n) and $c = $ l.c.m. $[m, n]$. Then $\mathbb{Q}(\zeta_m) \cap \mathbb{Q}(\zeta_n) = \mathbb{Q}(\zeta_d)$ and the composite field $\mathbb{Q}(\zeta_m)\mathbb{Q}(\zeta_n) = \mathbb{Q}(\zeta_c)$. In particular, the intersection and composite of cyclotomic fields are cyclotomic fields.

3. The group $\mu_m \subseteq \mathbb{Q}(\zeta_m)$. If σ is a $\mathbb{Q}$-isomorphism of $K = \mathbb{Q}(\zeta_m)$ into $\mathbb{C}$, then for ζ in μ_m, $(\sigma(\zeta))^m = \sigma(\zeta^m) = \sigma(1) = 1$, hence $\sigma(\zeta) \in K$. Thus K is a normal extension of $\mathbb{Q}$.

We now record this trivial but important fact.

Theorem 8.2. $\mathbb{Q}(\zeta_m)/\mathbb{Q}$ *is a Galois extension for all* $m \geq 1$.

Definition 8.3. The minimal polynomial $\Phi_m(x)$ of $\zeta_m = \cos\left(\frac{2\pi}{m}\right) + i\sin\left(\frac{2\pi}{m}\right)$ is called the m-th *cyclotomic polynomial*.

We shall show that all the primitive m-th roots of unity have $\Phi_m(x)$ as their minimal polynomial, i.e. the primitive roots in μ_m are the conjugates of ζ_m. We shall need the following calculation.

Theorem 8.4. *Suppose* ζ *is any primitive m-th root of unity. Then*

$$\prod_{\substack{1 \leq i,j \leq m \\ i \neq j}} (\zeta^i - \zeta^j) = (-1)^{m-1} m^m. \tag{8.2}$$

Proof. All the roots of $x^m - 1$ are $\zeta, \zeta^2, \ldots, \zeta^m = 1$. Hence

$$x^m - 1 = \prod_{j=1}^{m} (x - \zeta^j). \tag{8.3}$$

Equating the constant terms in (8.3), we get

$$(-1)^{m-1} = \prod_{j=1}^{m} \zeta^j, \tag{8.4}$$

whereas, on differentiating (8.3) and evaluating the resulting equation at $x = \zeta^i$ for $i = 1, \ldots, m$, we get

$$m\zeta^{i(m-1)} = \prod_{\substack{j=1 \\ j \neq i}}^{m} (\zeta^i - \zeta^j). \tag{8.5}$$

Now take the product for all $i = 1, \ldots, m$ to get

$$m^m \left(\prod_{i=1}^{m} \zeta^i \right)^{m-1} = \prod_{\substack{1 \leq i,j \leq m \\ i \neq j}} (\zeta^i - \zeta^j). \tag{8.6}$$

Substituting for $\prod_{i=1}^{m} \zeta^i$ from (8.4) in (8.6), we obtain (8.2). □

Corollary 8.5. *If $\zeta = \zeta_m$, $K = \mathbb{Q}(\zeta)$, then the discriminant $d_K | m^m$. In particular, if $m = p^r$ ($r \geq 1$) is a prime power, then $|d_K| = p^s$ ($s \geq 1$).*

Proof. Being minimal, d_K divides the discriminant of any basis of $K/\mathbb{Q}$, consisting of elements of $\mathcal{O}_K$. In particular, $d_K | \Delta(1, \zeta, \ldots, \zeta^{n-1})$, where $n = [K : \mathbb{Q}]$. But $\Delta(1, \zeta, \ldots, \zeta^{n-1})$ is the square of the van der Monde determinant $\det(\sigma_i(\zeta^j))$, where $\sigma_1, \ldots, \sigma_n$ are all the elements of $\mathrm{Gal}(K/\mathbb{Q})$. Hence,

$$\Delta(1, \zeta, \ldots, \zeta^{n-1}) = \prod_{i \neq j} (\sigma_i(\zeta) - \sigma_j(\zeta)).$$

This is a subproduct of the left-hand side of (8.2), which shows that $d_K | m^m$. □

Theorem 8.6. *Suppose $\Phi_m(x)$ is the minimal polynomial of ζ_m over $\mathbb{Q}$ and $K = \mathbb{Q}(\zeta_m)$. Then we have the following.*

 1. $\Phi_m(x) \in \mathbb{Z}[x]$,

 2. $\Phi_m(x) = \prod_{\substack{\zeta \in \mu_n, \\ \zeta \ primitive}} (x - \zeta).$
 In particular, $\deg \Phi_m(x) = \phi(m)$.

 3. $\mathrm{Gal}(K/\mathbb{Q}) \cong (\mathbb{Z}/m\mathbb{Z})^\times.$

Proof. (1) is just the well known Gauss' Lemma (see [23, p. 120]).

We shall prove (2) and (3) simultaneously. Suppose $\sigma \in G = \mathrm{Gal}(K/\mathbb{Q})$. Then σ is uniquely determined by the value $\sigma(\zeta_m)$. Moreover, $\zeta = \sigma(\zeta_m)$ is also a primitive m-th root of unity, for otherwise $\sigma^{-1}(\zeta) = \zeta_m$ would not be a primitive root of unity, either. So $\sigma(\zeta_m) = \zeta_m^a$, $1 \leq a = a(\sigma) \leq m$ with

$(a, m) = 1$. Further, it is easy to check that $a(\sigma_1) = a(\sigma_2) \Leftrightarrow \sigma_1 = \sigma_2$. Thus, the map

$$G \ni \sigma \to a(\sigma) \in (\mathbb{Z}/m\mathbb{Z})^\times \qquad (8.7)$$

is injective. In particular,

$$\deg \Phi_m(x) = |\operatorname{Gal}(\mathbb{Q}(\zeta)/\mathbb{Q})| \leq \phi(m). \qquad (8.8)$$

This map $\sigma \to a(\sigma)$ is easily seen to be a group homomorphism. If we show that $\Phi_m(x)$ vanishes at all the primitive m-th roots of unity, then we will have

$$\deg \Phi_m(x) \geq \phi(m),$$

which together with (8.7) and (8.8) will prove everything.

If ζ and η are two primitive m-th root of 1, then $\eta = \zeta^a$ with $(a, m) = 1$. Thus η may be obtained from ζ by replacing it successively with ζ^p, for each prime divisor p of a. Hence, it is enough to show that if a primitive root ζ of 1 is a zero of $\Phi_m(x)$, then so is ζ^p.

So fix a primitive m-th root ζ of 1 with $\Phi_m(\zeta) = 0$ and a prime p, not dividing m. First note by (1), i.e. by the lemma of Gauss, $\Phi_m(x)$ is a monic polynomial in $\mathbb{Z}[x]$. By the Multinomial Theorem,

$$(\Phi_m(x))^p - \Phi_m(x^p) \in p\mathbb{Z}[x].$$

But $\Phi_m(\zeta) = 0$. Therefore $\Phi_m(\zeta^p)$ is divisible in $\mathbb{Z}[\zeta]$ by p. Now write,

$$\Phi_m(x) = \prod_{\omega \in I} (x - \omega),$$

where I is a subset of μ_m. If $\Phi_m(\zeta^p) \neq 0$, then $\prod_{\omega \in I}(\zeta^p - \omega)$ is a subproduct of

$$\prod_{\substack{1 \leq i, j \leq m \\ i \neq j}} (\zeta^i - \zeta^j).$$

Hence, $\Phi_m(\zeta^p) | m$ in $\mathbb{Z}(\zeta)$. But $p | \Phi_m(\zeta^p)$ in $\mathbb{Z}[\zeta]$. Hence $p | m^m$ in $\mathbb{Z}[\zeta]$. But this can happen only if $p | m^m$ in $\mathbb{Z}$, which is a contradiction. $\qquad \square$

Theorem 8.7. *Suppose $m = p^r$ (p prime, $r \geq 1$), $\zeta = \zeta_m$ and $K = \mathbb{Q}(\zeta)$. Then p is the only prime ramified in K. Further, p is totally ramified in K, i.e. $(p) = \mathfrak{p}^{\phi(m)}$ for a prime ideal $\mathfrak{p}$ in K. The ideal $\mathfrak{p}$ is principal, generated by $1 - \zeta$.*

Proof. We know that p is ramified in K if and only if $p | d_K$ and by Corollary 8.5, $d_K = p^s$ ($s \geq 1$). Hence p is the only ramified prime.

To show that p is totally ramified, first note that ζ in μ_m, for $m = p^r$, is primitive if and only if $\zeta^{m/p} \neq 1$. Hence, the primitive m-th roots of 1 are precisely the roots of the polynomial $(x^m - 1)/(x^{m/p} - 1)$ in $\mathbb{Z}[x]$, i.e.

$$\frac{x^m - 1}{x^{m/p} - 1} = \prod_{\substack{1 \leq j \leq m, \\ (j,m)=1}} (x - \zeta^j).$$

This, together with L'Hospital's rule, gives

$$\prod_{\substack{1 \leq j \leq m \\ (j,m)=1}} (1 - \zeta^j) = \lim_{x \to 1} \prod_{\substack{1 \leq j \leq m \\ (j,m)=1}} (x - \zeta^j) = \lim_{x \to 1} \frac{x^m - 1}{x^{m/p} - 1} = p,$$

i.e.

$$p = \prod_{\substack{1 \leq j \leq m \\ (j,m)=1}} (1 - \zeta^j). \tag{8.9}$$

To compute the right hand side of (8.9), first note that $\frac{1-\zeta^j}{1-\zeta} = 1 + \zeta + \cdots + \zeta^{j-1} \in \mathcal{O}_K$. On the other hand, because $(j, m) = 1$, $ij \equiv 1 \pmod{m}$ for some i. Therefore the field element $\frac{1-\zeta}{1-\zeta^j} = \frac{1-\zeta^{ij}}{1-\zeta^j} = 1 + \zeta^j + \zeta^{2j} + \cdots + \zeta^{(i-1)j}$ is also in $\mathcal{O}_K$. This shows that $\frac{1-\zeta^j}{1-\zeta}$ and $\frac{1-\zeta}{1-\zeta^j}$ are both units in $\mathcal{O}_K$. Now returning to equation (8.9),

$$p = \prod_{\substack{1 \leq j \leq m \\ (j,m)=1}} (1 - \zeta^j) = (1 - \zeta)^{\phi(m)} \prod_{\substack{1 \leq j \leq m \\ (j,m)=1}} \frac{1 - \zeta^j}{1 - \zeta} = u(1 - \zeta)^{\phi(m)}$$

for u in $\mathcal{O}_K^\times$. Therefore $(p) = (1 - \zeta)^{\phi(m)}$. Since $\phi(m) = [\mathbb{Q}(\zeta) : \mathbb{Q}] = efg$, $f = g = 1$, $e = \phi(m)$ and $(1 - \zeta)$ has to be a prime ideal $\mathfrak{p}$. $\square$

Corollary 8.8. *Given a set* $S = \{p_1, \ldots, p_r\}$ *of distinct primes, there is an extension* $K/\mathbb{Q}$, *such that* p *ramifies in* K *if and only if* $p \in S$.

Proof. Take K to be the composite $K_1 \cdots K_r$ of $K_j = \mathbb{Q}(\zeta_{p_j})$. $\square$

Example 8.9. We illustrate Theorem 8.7 with two examples

(1) Let $m = 2^2$, so that $\zeta = \zeta_4 = e^{2\pi i/4} = i = \sqrt{-1}$. Then $K = \mathbb{Q}(\zeta) = \mathbb{Q}(i)$ and $[K : \mathbb{Q}] = \phi(m) = 2$. Further, $\mathcal{O}_K = \mathbb{Z} \oplus \mathbb{Z}i = \mathbb{Z}[i]$, the Gaussian integers. We have

$$d_K = \begin{vmatrix} 1 & i \\ 1 & -i \end{vmatrix}^2 = -4.$$

By Theorem 8.7, 2 is the only prime to ramify in K and $(2) = (1 - i)^2$. This can be checked directly: In $\mathbb{Z}[i]$, the ideal

$$(1 - i)^2 = ((1 - i)^2) = (-2i) = (2).$$

(2) Let $m = 3$, $\zeta = \omega = e^{2\pi i/3}$, $K = \mathbb{Q}(\omega)$, $[K : \mathbb{Q}] = \phi(3) = 2$. We have

$$(1 - \omega)^2 = ((1 - \omega)^2) = (1 + \omega^2 - 2\omega)$$
$$= (-\omega - 2\omega) = (-3\omega) = (3).$$

EXERCISES

1. Suppose $m = p^r$ (p odd, $r \geq 1$). Show that $\text{Gal}(\mathbb{Q}(\zeta_m)/\mathbb{Q})$ is cyclic of order $(p-1)p^{r-1}$.

2. If $m = 2^r$ ($r \geq 3$), show that $\text{Gal}(\mathbb{Q}(\zeta_m)/\mathbb{Q})$ is a direct product $G_1 \times G_2$, where $G_1 = \{1, \text{complex conjugation }\}$ and $|G_2| = p^{r-2}$.

Theorem 8.10. *Let $m > 2$ be an integer such that either m is odd or $4|m$. Then a prime p ramifies in $K = \mathbb{Q}(\zeta_m)$ if and only if $p|m$.*

Proof. If p is ramified, $p|d_K$. But by Corollary 8.5, d_K divides m^m. Hence $p|m$.

Conversely, let $p|m$. To show that p is ramified in K, it is enough to find a subextension k of K in which p ramifies. For this, we use Theorem 8.7.

Case (1). If p is odd, take $k = \mathbb{Q}(\zeta_p)$, and we are done.

Case (2). If $p = 2$, then $4|m$ and $\zeta_m^{m/4} = \sqrt{-1}$. We take $k = \mathbb{Q}(\sqrt{-1})$. Then 2 is ramified in k.   □

Remark 8.11. The converse is not true when $m = 2m'$, with m' odd. We leave it as an exercise to show that 2 is not ramified in $\mathbb{Q}(\zeta_6)$.

Theorem 8.12. *Let $K = \mathbb{Q}(\alpha)$ be a number field of degree n over $\mathbb{Q}$. The discriminant $\Delta(1, \alpha, \ldots, \alpha^{n-1})$ of the basis $1, \alpha, \ldots, \alpha^{n-1}$ of K over $\mathbb{Q}$ is given by*

$$\Delta(1, \alpha, \ldots, \alpha^{n-1}) = N_{K/\mathbb{Q}}(f'(\alpha)), \tag{8.10}$$

where $f(x)$ is the irreducible polynomial of α over $\mathbb{Q}$.

Proof. Let $G = \text{Gal}(\mathbb{Q}(\alpha)/\mathbb{Q}) = \{\sigma_1, \ldots, \sigma_n\}$. Then the discriminant $\Delta = \Delta(1, \alpha, \ldots, \alpha^{n-1})$ of $1, \alpha, \ldots, \alpha^{n-1}$ is the square of the van der Monde determinant

$$\begin{vmatrix} 1 & \cdots & 1 \\ \sigma_1(\alpha) & \cdots & \sigma_n(\alpha) \\ \vdots & & \\ \sigma_1(\alpha^{n-1}) & \cdots & \sigma_n(\alpha^{n-1}) \end{vmatrix} = \prod_{i<j}(\sigma_i(\alpha) - \sigma_j(\alpha)).$$

Hence

$$\Delta = \prod_{i \neq j} (\sigma_i(\alpha) - \sigma_j(\alpha))$$

$$= \prod_{i=1}^{n} \sigma_i \left(\prod_{j \neq i} (\alpha - \sigma_i^{-1}\sigma_j(\alpha)) \right)$$

$$= \prod_{i=1}^{n} \sigma_i \left(\prod_{\substack{\sigma \in G \\ \sigma \neq id}} (\alpha - \sigma(\alpha)) \right). \tag{8.11}$$

Now

$$f(x) = \prod_{j=1}^{n} (x - \sigma_j(\alpha)).$$

If we differentiate $f(x)$ and evaluate at $x = \alpha$, we get

$$f'(\alpha) = \prod_{\sigma \neq id} (\alpha - \sigma(\alpha)).$$

Hence by (8.11),

$$\Delta = \prod_{i=1}^{n} \sigma_i(f'(\alpha))$$

$$= N_{K/\mathbb{Q}}(f'(\alpha)). \qquad \square$$

Proposition 8.13. *If $q = p^r > 2$ is a prime power, $\zeta = \zeta_q$, then the discriminant*

$$\Delta(1, \zeta, \ldots, \zeta^{\phi(q)-1}) = (-1)^{\frac{\phi(q)}{2}} \frac{q^{\phi(q)}}{p^{q/p}}. \tag{8.12}$$

Proof. As before,

$$\Phi_q(x) = \frac{x^q - 1}{x^{q/p} - 1}$$

is in $\mathbb{Z}[x]$. Differentiate this expression for $\Phi_q(x)$ and evaluate at $x = \zeta$ to get

$$\Phi'_q(\zeta) = \frac{q\zeta^{q-1}}{\zeta^{q/p} - 1}.$$

On taking the norms of both sides, we get

$$N_{K/\mathbb{Q}}(\Phi'_q(\zeta)) = \frac{(-1)^{\frac{\phi(q)}{2}} q^{\phi(q)}}{N_{K/\mathbb{Q}}(\zeta^{q/p} - 1)}. \tag{8.13}$$

Since $\eta = \zeta^{q/p}$ is a primitive p-th root of unity,

$$N_{K/\mathbb{Q}}(\zeta^{q/p} - 1) = N_{K/\mathbb{Q}}(\eta - 1) = N_{\mathbb{Q}(\eta)/\mathbb{Q}}(N_{K/\mathbb{Q}(\eta)}(\eta - 1)),$$

i.e.

$$N_{K/\mathbb{Q}}(\zeta^{q/p} - 1) = (N_{\mathbb{Q}(\eta)/\mathbb{Q}}(\eta - 1))^{[K:\mathbb{Q}(\eta)]}. \tag{8.14}$$

But

$$[K : \mathbb{Q}(\eta)] = \frac{[K : \mathbb{Q}]}{[\mathbb{Q}(\eta) : \mathbb{Q}]} = \frac{\phi(p^r)}{\phi(p)} = \frac{q}{p} \tag{8.15}$$

and

$$N_{\mathbb{Q}(\eta)/\mathbb{Q}}(\eta - 1) = \prod_{a=1}^{p-1}(1 - \zeta_p^a) = \lim_{x \to 1} \Phi_p(x) = \lim_{x \to 1} \frac{x^p - 1}{x - 1} = p. \tag{8.16}$$

Hence, we get (8.12) from (8.10) and (8.13)–(8.16). $\square$

Proposition 8.14. *If $q = p^r > 2$ is a prime power, $\zeta = \zeta_q$ and $K = \mathbb{Q}(\zeta)$, then $\mathcal{O}_K = \mathbb{Z}[\zeta] = \mathbb{Z} \oplus \mathbb{Z}\zeta \oplus \cdots \oplus \mathbb{Z}\zeta^{\phi(q)-1}$.*

Proof. Clearly, $\mathbb{Z}[\zeta] \subseteq \mathcal{O}_K$. So we only need to show that $\mathcal{O}_K \subseteq \mathbb{Z}[\zeta]$. We shall use the fact that $\omega = 1 - \zeta$ is a uniformizing parameter at $\mathfrak{p}$, where $(p) = \mathfrak{p}^{\phi(q)}$.

It is obvious that $\mathbb{Z}[\zeta] = \mathbb{Z}[\omega]$. Hence it suffices to show that $\mathcal{O}_K \subseteq \mathbb{Z}[\omega]$. Since $K = \mathbb{Q}(\zeta) = \mathbb{Q}(\omega)$, it follows that $1, \omega, \ldots, \omega^{\phi(q)-1}$ is also a basis of K over $\mathbb{Q}$. Therefore, if $\alpha \in K$, we have

$$\alpha = \sum_{i=0}^{\phi(q)-1} a_i \omega^i = \sum_{j=0}^{\phi(q)-1} b_j \zeta^j$$

with a_j, b_j in $\mathbb{Q}$. We need to show that $a_j \in \mathbb{Z}$.

Let $G = \mathrm{Gal}(\mathbb{Q}(\zeta)/\mathbb{Q}) = \{\sigma_1, \ldots, \sigma_n\}$ with $n = \phi(q)$. Then

$$\sigma_i(\alpha) = \sum_{j=0}^{n-1} b_j \sigma_i(\zeta^j), \; i = 1, \ldots, n. \tag{8.17}$$

Solving the system (8.17) of n linear equations in n variables, $b_0, \ldots, b_{n-1}$ by Cramer's rule, we see that $b_j = \frac{\gamma_j}{\Delta}$, where $\gamma_j \in \mathcal{O}_K$ and $\Delta = \Delta(1, \zeta, \ldots, \zeta^{n-1})$. By Proposition 8.13, Δ is a power of p. Therefore, for all j, $p^m b_j \in \mathbb{Z}$ and hence, $p^m a_j \in \mathbb{Z}$, if m is sufficiently large. Let $m \geq 0$ be the smallest integer such that $p^m a_j \in \mathbb{Z}$ for all j. We need to show that $m = 0$. Suppose $m \geq 1$. Then there is an index i, $0 \leq i \leq n - 1$ such that $p^m a_j \in p\mathbb{Z}$ for $j = 0, 1, \ldots, i - 1$ but $p^m a_i \notin p\mathbb{Z}$.

Since $m > 0$, $p^m \alpha \in p\mathcal{O}_K = \omega^n \mathcal{O}$. Also for $j = 0, 1, \ldots, i-1$, $p^m a_j \in p\mathbb{Z} \subseteq \omega^n \mathcal{O}_K$. Hence

$$\beta = \sum_{j=i}^{n-1} p^m a_j \omega^j = p^m \alpha - \sum_{j=0}^{i-1} p^m a_j \omega^j$$

is in $\omega^n \mathcal{O}_K$. (If $i = 0$, the second sum is absent.) Thus

$$p^m a_j \omega^i = \beta - \sum_{j=i+1}^{n-1} p^m a_j \omega^{j-1}$$

is in $\omega^{i+j} \mathcal{O}_K$. Hence $p^m a_i$ is in $\omega \mathcal{O} \cap \mathbb{Z} = p\mathbb{Z}$. This contradicts the choice of i. $\qquad\qquad\square$

Finally, we are in a position to compute the discriminant of an arbitrary cyclotomic field. Moreover, we shall show that if $m > 2$ is arbitrary, $\zeta = \zeta_m$ and $K = \mathbb{Q}(\zeta)$, then $\mathcal{O}_K = \mathbb{Z}[\zeta]$.

In the remaining section, unless stated to the contrary, let K_1, K_2 be two number fields such that $[K : \mathbb{Q}] = [K_1 : \mathbb{Q}][K_2 : \mathbb{Q}]$, where $K = K_1 K_2$. We denote by d_j the discriminant of K_j and by $\mathcal{O}_j$, the ring of integers of K_j ($j = 1, 2$). We recall (without proof) the following fact from Galois theory.

Proposition 8.15. *Given $\mathbb{Q}$-isomorphisms $\sigma_j : K_j \to \mathbb{C}$ ($j = 1, 2$), there is a $\mathbb{Q}$-isomorphism $\sigma : K \to \mathbb{C}$, such that $\sigma_{|K_j} = \sigma_j$.*

We denote by $\mathcal{O}_1 \mathcal{O}_2$ the smallest subring of K containing both $\mathcal{O}_1$ and $\mathcal{O}_2$. It consists of all the finite sums of the form

$$x_1 y_1 + \ldots + x_N y_N \ (x_j \in \mathcal{O}_1, y_j \in \mathcal{O}_2).$$

We shall need the following.

Proposition 8.16. *Let a be the g.c.d of the discriminants d_1, d_2 of K_1, K_2, respectively. Then $a\mathcal{O}_K \subseteq \mathcal{O}_1 \mathcal{O}_2$.*

Proof. Let $\mathcal{O}_1 = \mathbb{Z}\alpha_1 \oplus \ldots \oplus \mathbb{Z}\alpha_m$ and $\mathcal{O}_2 = \mathbb{Z}\beta_1 \oplus \ldots \oplus \mathbb{Z}\beta_n$ with $d_1 = \Delta(\alpha_1, \ldots, \alpha_m)$ and $d_2 = \Delta(\beta_1, \ldots, \beta_n)$.

Since $\alpha_1, \ldots, \alpha_m$ is a basis of K_1 over $\mathbb{Q}$, and $\beta_1, \ldots, \beta_n$ is a basis of K_2 over $\mathbb{Q}$, the condition $[K_1 K_2 : \mathbb{Q}] = [K_1 : \mathbb{Q}][K_2 : \mathbb{Q}]$ implies that $\{\alpha_i \beta_j\}$ is a basis of $K_1 K_2$ over $\mathbb{Q}$. If $\alpha \in \mathcal{O}_K$, then surely,

$$\alpha = \sum_{i,j} \frac{a_{ij}}{b} \alpha_i \beta_j \tag{8.18}$$

with a_{ij}, b in $\mathbb{Z}$ and $b > 0$. We take $b > 0$ to the smallest common denominator of the coefficients of $\alpha_i \beta_j$. Then b is coprime to at least one a_{ij}. We need to show that for any α in $\mathcal{O}_K$, this least common denominator $b | a = \gcd(d_1, d_2)$.

Let $\sigma_1, \ldots, \sigma_m$ be the m $\mathbb{Q}$-isomorphisms of K_1 into $\mathbb{C}$. By Proposition 8.15, each σ can be extended to a $\mathbb{Q}$-isomorphism of K into $\mathbb{C}$, whose restriction to K_2 is the identity of K_2. We denote this extension by σ also. We apply $\sigma = \sigma_r$ ($r = 1, \ldots, m$) to (8.18) to get

$$\sigma_r(\alpha) = \sum_{i=1}^{m} \left(\sum_{j=1}^{n} \frac{a_{ij}\beta_j}{b} \right) \sigma_r(\alpha_i). \qquad (8.19)$$

We put

$$c_i = \sum_{j=1}^{n} \frac{a_{ij}}{b} \beta_j.$$

Then the system of equation (8.19) becomes

$$\sum_{i=1}^{n} \sigma_r(\alpha_i) c_i = \sigma_r(\alpha), \quad (r = 1, \ldots, m). \qquad (8.20)$$

Solving (8.20) by Cramer's rule, we get $c_i = \frac{\gamma_i}{\Delta}$, where $\gamma_i \in \mathcal{O}_K$ and the rational integer

$$\Delta = \Delta(\alpha_1, \ldots, \alpha_m) = d_1.$$

This shows that for $i = 1, \ldots, m$,

$$d_1 c_i = \sum_{j=1}^{n} d_1 \frac{a_{ij}}{b} \beta_j$$

is in $\mathcal{O}_K$ as well as in K_2, and hence in $\mathcal{O}_2 = \mathcal{O}_K \cap K_2$. But $\mathcal{O}_2 = \mathbb{Z}\beta_1 \oplus \cdots \oplus \mathbb{Z}\beta_n$. Hence, $d_1 \frac{a_{ij}}{b} \in \mathbb{Z}$, for all i, j. Since b is coprime to at least one a_{ij}, $b | d_1$. Similarly, $b | d_2$. Hence $b | a = \gcd(d_1, d_2)$. $\qquad \square$

Theorem 8.17 (Integral Basis for Cyclotomic Fields). *Suppose $m > 2$ is any integer, $\zeta = \zeta_m$ and $K = \mathbb{Q}(\zeta)$. Then*

$$\mathcal{O}_K = \mathbb{Z}[\zeta] = \mathbb{Z} \oplus \mathbb{Z}\zeta \oplus \cdots \oplus \mathbb{Z}\zeta^{\phi(m)-1}.$$

Proof. We use induction on the number s of prime divisors of m. If $s = 1$, the theorem has already been proved as Proposition 8.14.

If $s > 1$, let $m_1 = p^r$ be the largest power of a prime p appearing in m. Let $m_2 = m/m_1$. Then $m = m_1 m_2$ with $m_1, m_2 > 1$ and $\gcd(m_1, m_2) = 1$. For $j = 1, 2$, let $K_j = \mathbb{Q}(\zeta_{m_j})$, $d_j = d_{K_j}$. By Corollary 8.5, $\gcd(d_1, d_2) = 1$.

By induction hypothesis on s, $\mathcal{O}_j = \mathcal{O}_{K_j} = \mathbb{Z}[\zeta_{m_j}]$. Since $K = \mathbb{Q}(\zeta_m) = K_1 K_2$ and $a = \gcd(d_1, d_2) = 1$, by Proposition 8.16, $\mathcal{O}_K \subseteq \mathcal{O}_1 \mathcal{O}_2 = \mathbb{Z}[\zeta_m]$. Because the reverse inclusion, $\mathbb{Z}[\zeta_m] \subseteq \mathcal{O}_K$, is obvious, $\mathcal{O}_K = \mathbb{Z}[\zeta_m]$. $\qquad \square$

Corollary 8.18. *If $m > 1$ is any integer and $K = \mathbb{Q}(\zeta_m)$, then the discriminant*

$$d_K = (-1)^{\phi(m)/2} \cdot m^{\phi(m)} / \prod_{p|m} p^{\phi(m)/(p-1)}.$$

Proof. Let $n = \phi(m)$ and $G = \mathrm{Gal}(K/\mathbb{Q}) = \{\sigma_1, \ldots, \sigma_n\}$, with $\sigma_1 = id$. If $\zeta = \zeta_m$, let $\omega_1 = \zeta$, $\omega_2 = \sigma_2(\zeta), \ldots, \omega_n = \sigma_n(\zeta)$ be the n conjugates of ζ. Since $\{1, \zeta, \ldots, \zeta^{n-1}\}$ is a $\mathbb{Z}$-basis of $\mathcal{O}_K$, the discriminant d_K is the square of the van der Monde determinant

$$\begin{vmatrix} 1 & \cdots & 1 \\ \sigma_1(\zeta) & \cdots & \sigma_n(\zeta) \\ \vdots & & \\ \sigma_1(\zeta^{n-1}) & \cdots & \sigma_n(\zeta^{n-1}) \end{vmatrix} = \begin{vmatrix} 1 & \omega_1 & \cdots & \omega_1^{n-1} \\ 1 & \omega_2 & \cdots & \omega_2^{n-1} \\ \vdots & & & \\ 1 & \omega_n & \cdots & \omega_n^{n-1} \end{vmatrix} = \prod_{i<j}(\omega_i - \omega_j).$$

To complete the proof, we leave it as an exercise to compute the product

$$\prod_{i<j}(\omega_i - \omega_j),$$

where $\omega_1, \ldots, \omega_n$ are all the primitive m-th roots of unity. $\qquad\square$

EXERCISE

(Kummer's Lemma) Suppose $m > 1$ and $K = \mathbb{Q}(\zeta_m)$. If for u in $\mathbb{C}$, $\overline{u}$ denotes its complex conjugate, show that $u \in \mathcal{O}_K^\times$ implies that $u/\overline{u} \in \mu_m$, the group of m-th roots of unity.

 Hint: Let $\zeta = \zeta_m$. Given σ in $\mathrm{Gal}(K/\mathbb{Q})$, $\sigma(\zeta) = \zeta^a$ for some a. Since $\overline{\sigma(\zeta)} = \overline{\zeta^a} = \overline{\zeta}^a$, σ commutes with complex conjugation. Therefore, if $\alpha = u/\overline{u}$, then $\alpha \in \mathcal{O}_K^\times$ and $|\sigma(\alpha)| = 1$ for all σ in G. Hence, α is a root of unity in K.

8.2 Arithmetic in Cyclotomic Fields

Let $m > 2$, $\zeta = \zeta_m$ and $K = \mathbb{Q}(\zeta)$ be fixed. We now consider an arbitrary prime p with $\gcd(p, m) = 1$. Then the Artin symbol $\left(\frac{K/\mathbb{Q}}{p}\right)$ is the unique element σ of the Galois group $G = \mathrm{Gal}(K/\mathbb{Q})$ characterized by

$$\sigma(\alpha) \equiv \alpha^p \pmod{\mathfrak{P}}, \tag{8.21}$$

$\forall\, \alpha$ in $\mathcal{O}_K$. In equation (8.21), $\mathfrak{P}$ is any prime of K dividing p. On the other hand, σ is also uniquely determined by the value of $\sigma(\zeta)$, which is a primitive m-th root of unity and hence ζ^a for some $a \in \mathbb{N}$ with $(a, m) = 1$. Since $\mathcal{O}_K = \mathbb{Z}[\zeta]$ and $\forall\, a$ in $\mathbb{Z}$, $a^p \equiv a \pmod{p}$, condition (8.21) is equivalent to

$$\zeta^a \equiv \zeta^p \pmod{\mathfrak{P}}. \tag{8.22}$$

Furthermore, it follows at once from the following Proposition that the condition (8.22) is further equivalent to

$$a \equiv p \pmod{m}. \tag{8.23}$$

Proposition 8.19. *Suppose p and $\mathfrak{P}$ are as above. Let $a, b \in \mathbb{N}$ with $1 \le a, b \le m$, $(a, m) = (b, m) = 1$. If*

$$\zeta^a \equiv \zeta^b \pmod{\mathfrak{P}},$$

then $a = b$.

Proof. Suppose $a \ge b$. Then $\zeta^a \equiv \zeta^b \pmod{\mathfrak{P}} \Leftrightarrow \zeta^{a-b} \equiv 1 \pmod{\mathfrak{P}}$. Hence, it is enough to show that if $0 \le a < m-1$, then $\zeta^a \equiv 1 \pmod{\mathfrak{P}} \Leftrightarrow a = 0$. The implication $\Leftarrow$ is trivial. To show that $\zeta^a \equiv 1 \pmod{\mathfrak{P}} \Rightarrow a = 0$, differentiate

$$x^m - 1 = \prod_{j=0}^{m-1} (x - \zeta^j)$$

and evaluate at $x = 1$ to get

$$m = \prod_{j=1}^{m-1} (1 - \zeta^j).$$

Hence, if $a > 0$, then $\zeta^a \equiv 1 \pmod{\mathfrak{P}}$ implies that $\mathfrak{P}|m$, which implies that $p|m$. This contradicts our assumption that $\gcd(p, m) = 1$. $\qquad\square$

We summarize this discussion as the following important fact.

Theorem 8.20 (Cyclotomic Reciprocity Law). *The Artin symbol for the cyclotomic field $\mathbb{Q}(\zeta_m)$ is given by*

$$\left(\frac{\mathbb{Q}(\zeta_m)/\mathbb{Q}}{p} \right) = \zeta_m^p,$$

where the prime p is coprime to m.

Corollary 8.21. *Suppose p is any prime with $(p, m) = 1$. Choose the smallest integer $f \ge 1$ such that $p^f \equiv 1 \pmod{m}$. Let $\mathfrak{P}$ be any prime in $K = \mathbb{Q}(\zeta_m)$ dividing p. Then $f = f_p = f(\mathfrak{P}/p)$, and p factors into $\phi(m)/f$ primes in K.*

Proof. The condition $(p, m) = 1$ implies that the inertia group $T_p = T_{\mathfrak{P}/p} = \{1\}$ and the decomposition group

$$Z_p = Z_{\mathfrak{P}/p} \cong \mathrm{Gal}((\mathcal{O}/\mathfrak{P})/(\mathbb{Z}/p\mathbb{Z})),$$

which is a cyclic group of order f, generated by the Frobenius automorphism Φ_p of $\mathcal{O}/\mathfrak{P}$ over $\mathbb{F}_p = \mathbb{Z}/p\mathbb{Z}$. Hence, its order is the smallest $f \geq 1$, such that $\Phi_p^f = id$, which is the case if and only if $\Phi_p(\zeta) = \zeta^{p^f} = \zeta \Leftrightarrow p^f \equiv 1 \pmod{m}$. Hence $f = f_p$. Since $e_p = 1$, we have $\phi(m) = f_p g_p$ and we are done. $\square$

Corollary 8.22. *A prime p, coprime to m, splits completely in $\mathbb{Q}(\zeta_m)$ if and only if $p \equiv 1 \pmod{m}$.*

Proof. The prime p with $(p, m) = 1$ splits completely in $K = \mathbb{Q}(\zeta_m) \Leftrightarrow$ the Artin symbol $\left(\frac{\mathbb{Q}(\zeta_m)/\mathbb{Q}}{p} \right) = 1 \Leftrightarrow \zeta^p = 1 \Leftrightarrow p \equiv 1 \pmod{m}$. $\square$

Corollary 8.23. *Let p_1, p_2 be two primes, both coprime to m, such that $p_1 \equiv p_2 \pmod{m}$. Then p_1, p_2 split into the same number of distinct prime factors in K.*

We may say that p_1, p_2 have the same splitting type in K.

Proof. Let $\mathfrak{P}_j$ be any prime divisor of p_j in K, $f_j \geq 1$ the smallest integer with $p^{f_j} \equiv 1 \pmod{m}$ for $j = 1, 2$. It is given that $p_1 \equiv p_2 \pmod{m}$. Hence $f_1 = f_2$, which implies that $g_1 - \frac{\phi(m)}{f_1} - \frac{\phi(m)}{f_2} = g_2$. $\square$

Example 8.24. We take $m = 10$, $p = 7$. The smallest $f \geq 1$ with $7^f \equiv 1 \pmod{10}$ is $f = 4$. Hence by Corollary 8.21, in $\mathbb{Q}(\zeta_{10})$, the prime 7 splits into $\phi(10)/4 = 4$ primes. The same is true, by Corollary 8.23, for all the primes $17, 37, 47, 67, 97, \ldots$

Remark 8.25. By Dirichlet's theorem on infinitude of primes in arithmetic progression, Corollary 8.23 implies the following general statement.

Theorem 8.26. *Given $m > 2$ and a prime p with $(p, m) = 1$, there are infinitely many primes, which split into the same number of primes in $\mathbb{Q}(\zeta_m)$ as p does.*

In particular, there are infinitely many primes p (namely, those with $p \equiv 1 \pmod{m}$) which split completely in $\mathbb{Q}(\zeta_m)$.

EXERCISES

1. Prove Remark 8.11.

2. Complete the proof of Corollary 8.18.

3. Specialize Theorem 6.19 to the cyclotomoic field $\mathbb{Q}(\zeta_m)$.

9

The Kronecker-Weber Theorem

This famous theorem asserts that every Abelian extension K of $\mathbb{Q}$ is a subfield of $\mathbb{Q}(\zeta_m)$ for some m. These days, it is customary to obtain the Kronecker-Weber Theorem as a corollary to the main theorems of class field theory. However, we shall give an elementary proof, based essentially on that of Hilbert.

9.1 Gauss Sums

We begin with a modest goal, namely, to show that every quadratic extension $K = \mathbb{Q}(\sqrt{d})$ with $\mathrm{Gal}(K/\mathbb{Q}) \cong \{\pm 1\}$, which is Abelian, is cyclotomic, i.e. contained in $\mathbb{Q}(\zeta_m)$ for some m. The main tool is the Gauss sum.

Let p be an odd prime and $\zeta = \zeta_p = e^{2\pi i/p}$. Recall the *Legendre symbol* $\left(\frac{a}{p}\right)$ for an integer a with $\gcd(a, p) = 1$. We define $\left(\frac{a}{p}\right) = 1$ or -1 according as the congruence $x^2 \equiv a \pmod{p}$ has a solution or has no solution. In other words

$$\left(\frac{a}{p}\right) = \begin{cases} 1 & \text{if } a \in \mathbb{F}_p^{\times 2} \\ -1 & \text{otherwise.} \end{cases}$$

It is convenient to set $\left(\frac{a}{p}\right) = 0$ if $p|a$. Since ζ^a depends only on the residue class of $a \pmod{p}$, ζ^a is well defined for a in $\mathbb{F}_p$.

Definition 9.1. Suppose $p > 2$ is a prime. The element $\gamma = \gamma_p$ of $\mathbb{Q}(\zeta_p)$, defined by

$$\gamma = \sum_{x \in \mathbb{F}_p} \left(\frac{x}{p}\right) \zeta^x \tag{9.1}$$

is called a *Gauss sum*.

Theorem 9.2. *For the Gauss sum $\gamma = \gamma_p$, we have*

$$\gamma^2 = \left(\frac{-1}{p}\right) p. \tag{9.2}$$

Proof. Since $\left(\frac{0}{p}\right) = 0$ for the zero element of $\mathbb{F}_p$,

$$\gamma^2 = \left(\sum_{x \in \mathbb{F}_p^\times} \left(\frac{x}{p}\right) \zeta^x\right) \left(\sum_{y \in \mathbb{F}_p^\times} \left(\frac{y}{p}\right) \zeta^y\right) = \sum_{x,y \in \mathbb{F}_p^\times} \left(\frac{xy}{p}\right) \zeta^{x+y}. \qquad (9.3)$$

For a fixed x in $\mathbb{F}_p^\times$, the map $\mathbb{F}_p^\times \ni y \to xy \in \mathbb{F}_p^\times$ permutes the elements of $\mathbb{F}_p^\times$. Hence in (9.3), we can replace the sum over y by the sum over xy to get

$$\gamma^2 = \sum_{x,y \in \mathbb{F}_p^\times} \left(\frac{x^2 y}{p}\right) \zeta^{x+xy} = \sum_{x,y \in \mathbb{F}_p^\times} \left(\frac{y}{p}\right) \zeta^{x(1+y)} \qquad (9.4)$$

$$= \sum_{\substack{x,y \in \mathbb{F}_p^\times \\ y \neq -1}} \left(\frac{y}{p}\right) \zeta^{x(1+y)} + \left(\frac{-1}{p}\right)(p-1).$$

Because the sum

$$1 + \zeta + \cdots + \zeta^{p-1} = 0,$$

for a fixed $y \neq -1$, the sum $\sum_{x \in \mathbb{F}_p^\times} \zeta^{x(1+y)}$ is, up to rearrangement of terms, equal to $\zeta + \zeta^2 + \cdots + \zeta^{p-1} = -1$. Hence,

$$\sum_{x,y \in \mathbb{F}_p^\times,\, y \neq -1} \left(\frac{y}{p}\right) \zeta^{x(1+y)} = \sum_{y \in \mathbb{F}_p^\times,\, y \neq -1} \left(\frac{y}{p}\right) \sum_{x \in \mathbb{F}_p^\times} \zeta^{x(1+y)}$$

$$= \left(\frac{-1}{p}\right) p - \sum_{y \in \mathbb{F}_p^\times} \left(\frac{y}{p}\right) = \left(\frac{-1}{p}\right) p.$$

The last sum is zero, because exactly half the elements of $\mathbb{F}_p^\times$ ($p > 2$) are squares. Hence (9.4) reduces to (9.2). $\qquad \square$

It is convenient to introduce the following terminology.

Definition 9.3. A number field K is called a *cyclotomic field* if $K \subseteq \mathbb{Q}(\zeta_m)$ for some $m \geq 1$.

Note that the composition of cyclotomic fields is a cyclotomic field.

Corollary 9.4. *The quadratic field $K = \mathbb{Q}(\sqrt{d})$ is a cyclotomic field.*

Proof. First, note that

1. $\sqrt{2} \in \mathbb{Q}(\zeta_8)$,
2. $\sqrt{-1} \in \mathbb{Q}(\zeta_4)$,

3. if p is an odd prime, then by Theorem 9.2, $\sqrt{p} \in \mathbb{Q}(\sqrt{-1}, \sqrt{p}) = \mathbb{Q}(\zeta_{4p})$.

Hence, if we write the square-free integer $d \neq 0, 1$ as

$$d = \pm 2^a p_1 \dots p_r \quad (r \geq 0, a = 0, 1),$$

where $p_1, \dots, p_r$ are distinct odd primes, then $\mathbb{Q}(\sqrt{d}) \subseteq \mathbb{Q}(\zeta_{8m})$, with $m = p_1 \cdots p_r$. $\qquad\square$

9.2   Proof of the Kronecker-Weber Theorem

We carry out the proof in a series of propositions and then put them together. We basically follow [19].

Proposition 9.5. *If 2 is the only prime to ramify in a cyclic extension $K/\mathbb{Q}$ of degree 2^m, then K is cyclotomic.*

For the sake of clarity, we break the proof into simple lemmas.

Lemma 9.6. *Suppose $K \subseteq \mathbb{R}$ is a cyclic extension of $\mathbb{Q}$ of degree 2^m ($m \geq 1$). If 2 is the only prime to ramify in K, then $\mathbb{Q}(\sqrt{2})$ is the unique quadratic subfield of K.*

Proof. Since $K/\mathbb{Q}$ is cyclic of degree 2^m ($m \geq 1$), it certainly contains a quadratic subfield $L = \mathbb{Q}(\sqrt{d})$, where $d \neq 0, 1$ is square-free. It is the fixed field of a subgroup of index two in $\mathrm{Gal}(K/\mathbb{Q})$. By Minkowski's theorem on discriminants, at least one prime, which can only be 2, must ramify in L. Hence 2 is the only prime divisor of its discriminant d_L. Thus $d = -1$ or 2. But $L \subseteq \mathbb{R}$. Hence $L = \mathbb{Q}(\sqrt{2})$. $\qquad\square$

Lemma 9.7. *Suppose $K \subseteq \mathbb{R}$ is an Abelian extension of $\mathbb{Q}$ of degree 2^m. If 2 is the only prime that can ramify in K, then $K/\mathbb{Q}$ is cyclic.*

Proof. For $m = 0, 1$, there is nothing to prove. We need to prove the lemma only for $m \geq 2$. If $K/\mathbb{Q}$ is not cyclic, then the Galois group $G = \mathrm{Gal}(K/\mathbb{Q})$ is a product

$$G = C_1 \times \cdots \times C_r \quad (r \geq 2)$$

of nontrivial cyclic groups C_j of order 2^{m_j}. We show that this leads to a contradiction. For simplicity, we take $r = 2$. (For $r > 2$, the proof is similar.) If K_j is the fixed field of C_j, then by Lemma 9.6, we have the Hasse diagram (cf. Figure 9.1) with $K = K_1 K_2$.

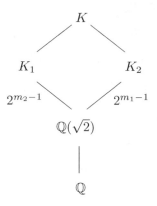

FIGURE 9.1: Hasse diagram for $K = K_1 K_2$.

Since $m = m_1 + m_2$, the degree $[K : \mathbb{Q}] = [K_1 K_2 : \mathbb{Q}] \leq 2^{m_1 + m_2 - 1} = 2^{m-1}$. This is a contradiction. $\qquad \square$

Lemma 9.8. *Suppose $K/\mathbb{Q}$ is an Abelian extension of degree 2^m, in which 2 is the only prime that can ramify. Then K is the composite of K_1, K_2, where $K_1 \subseteq \mathbb{R}$ is cyclic and $K_2 = \mathbb{Q}$ or $\mathbb{Q}(\sqrt{i})$.*

Proof. If $K \subseteq \mathbb{R}$, we are done (Lemma 9.7). If not, the complex conjugation τ is an element of the Galois group $G = \mathrm{Gal}(K/\mathbb{Q})$ of order two and

$$G = \langle \tau \rangle \times H,$$

where H is a subgroup of G of order 2^{m-1}. The fixed field $K_2 = K^H$ of H is an imaginary quadratic field in which 2 is the only prime to ramify. Hence $K_2 = \mathbb{Q}(i)$. The fixed field K_1 of τ is contained in $\mathbb{R}$ and is of degree 2^{m-1}. Again 2 is the only prime that can ramify in K_1. Hence by Lemma 9.7, $K_1/\mathbb{Q}$ is cyclic. $\qquad \square$

Proof of Proposition 9.5. We use induction on m. If $m = 1$, K is a quadratic field, hence it is cyclotomic. So suppose $m > 1$. First note that K contains a real cyclic subextension of $\mathbb{Q}$ of degree at least 2^{m-1}. It is the fixed field of complex conjugation. Hence by Lemma 9.6, $K \supseteq \mathbb{Q}(\sqrt{2})$.

Now let $n = 2^m$ and consider the subfield $L = \mathbb{Q}(\zeta_{4n} + \overline{\zeta}_{4n})$ of $\mathbb{Q}(\zeta_{4n})$. Since $L \subseteq \mathbb{R}$ and 2 is the only prime to ramify in L, by Lemma 9.7, $L/\mathbb{Q}$ is cyclic. The degree $[L : \mathbb{Q}]$ of the cyclic extension $L/\mathbb{Q}$ is $n = 2^m$. Hence by Lemma 9.6, $L \supseteq \mathbb{Q}(\sqrt{2})$. It is now clear from the Hasse diagram (Figure 9.2) that the degree of the composition KL over $\mathbb{Q}$ satisfies the inequality

$$[KL : \mathbb{Q}] < n^2. \tag{9.5}$$

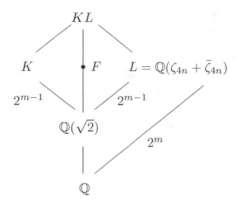

FIGURE 9.2: Hasse diagram for KL.

By Proposition 7.17, the Galois group $\Gamma = \mathrm{Gal}(KL/\mathbb{Q})$ is a proper subgroup of $G \times H$, where $G = \mathrm{Gal}(K/\mathbb{Q})$, $H = \mathrm{Gal}(L/\mathbb{Q})$. Let σ, τ be the generators of the cyclic groups G and H, respectively, which agree on $K \cap L$, so that $(\sigma, \tau) \in \Gamma$. The group Δ, generated by (σ, τ) has order n. Hence it follows from inequality (9.5) that the degree of the fixed field $F = (KL)^{\Delta}$ of Δ over $\mathbb{Q}$ satisfies the inequality

$$[F : \mathbb{Q}] \leq 2^{m-1}.$$

The prime 2 is still the only prime to ramify in F. By Lemmas 9.7 and 9.8 and the induction hypothesis on m, F is cylotomic, and therefore so is FL. Thus we will be done if we can show that $FL = KL$, since $KL \supseteq K$. But this is obvious, because only the identity automorphism in Δ restricts to the identity on L. $\qquad\square$

Proposition 9.9. *Suppose p is an odd prime and K is a cyclic extension of $\mathbb{Q}$ of degree p^m. If p is the only prime to ramify in K, then K is cyclotomic.*

Proof. Let $n = p^{m+1}$. The n-th cyclotomic field $\mathbb{Q}(\zeta_n)$ has degree $(p-1)p^m$ over $\mathbb{Q}$. Let L be the unique subfield of $\mathbb{Q}(\zeta_n)$ of degree p^m over $\mathbb{Q}$. We show that $K = L$. If not, then the compositum KL is Abelian of degree more than p^m over $\mathbb{Q}$, and in KL, p is the only prime that ramifies. Hence by Theorem 7.24, KL is cyclic. But (by Proposition 7.17) $\mathrm{Gal}(KL/\mathbb{Q})$ has no element of order larger than p^m. This contradiction proves that $K = L$. $\qquad\square$

In the next proposition, we remove the condition that l is the only prime to ramify in a cyclic extension of degree l^m ($l \geq 2$, a prime) for the field to be cyclotomic.

Proposition 9.10. *Let $l \geq 2$ be a prime and $K/\mathbb{Q}$ a cyclic extension of degree l^m $(m \geq 1)$. Then K is cyclotomic.*

Proof. We use induction on the number r of distinct primes $p_1, \ldots, p_r \neq l$ that ramify in K. If $r = 0$, we are already done (Propositions 9.5 and 9.9). So suppose $r \geq 1$. Let $p = p_r$ and $\mathfrak{P}$ a prime divisor of p in K. All subgroups and quotients of $G = \text{Gal}(K/\mathbb{Q})$ are cyclic and their orders are power of l. Since $(p, l) = 1$, by Theorem 7.10, all the higher ramification groups of $\mathfrak{P}/p$ are trivial and by Theorem 7.16, the order l^a of the inertia group $T = T_{\mathfrak{P}/p}$, i.e.

$$|T| = \ell^a,$$

divides $p - 1$. On the other hand, the cyclic extension $\mathbb{Q}(\zeta_p)/\mathbb{Q}$ of order $p - 1$ contains a unique subfield L with

$$[L : \mathbb{Q}] = l^a,$$

and p is the only prime to ramify in L. (In fact, p is totally ramified in L.) So, the degree of the extension $KL/\mathbb{Q}$ is (cf. Figure 9.3)

$$[KL : \mathbb{Q}] = l^{m+b} \ (b \leq a).$$

If $H = \text{Gal}(L/\mathbb{Q})$, then the Galois group $\text{Gal}(KL/\mathbb{Q})$ is isomorphic to a subgroup of $G \times H$. In fact the isomorphism is given by

$$\text{Gal}(KL/\mathbb{Q}) \ni \tau \to (\tau_{|K}, \tau_{|L}) \in G \times H.$$

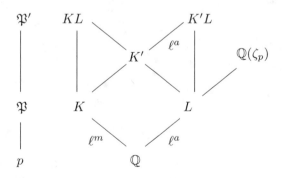

FIGURE 9.3: Hasse diagram.

If $\mathfrak{P}'$ is a prime divisor of $\mathfrak{P}$ in KL and $T' = T_{\mathfrak{P}'/p} \subseteq \text{Gal}(KL/\mathbb{Q})$ is the inertia group, we first show that the order

$$|T'| = l^a. \tag{9.6}$$

Put $\mathfrak{P}_L = \mathfrak{P} \cap L$. Then the order $|T'|$ is the ramification index $e(\mathfrak{P}'/p)$ and $e(\mathfrak{P}'/p) \geq e(\mathfrak{P}_L/p) = p^a$, because p is totally ramified in L and $[L : \mathbb{Q}] = p^a$. So

$$|T'| \geq l^a. \tag{9.7}$$

The restriction of τ in T' to K maps T' into T, so $T' \subseteq T \times H$. The higher ramification groups of $\mathfrak{P}'/p$ are trivial, so being isomorphic to a subgroup of the cyclic group $(\mathcal{O}_{KL}/\mathfrak{P}')^\times$, $T' \subseteq T \times H$ is also cyclic. Because $|T| = |H| = l^a$, no element of $T \times H$ has order larger than l^a. Hence

$$|T'| \leq l^a. \tag{9.8}$$

Now (9.6) follows from (9.7) and (9.8).

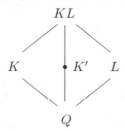

FIGURE 9.4: Another Hasse diagram.

Let K' be the fixed field of $T' \subseteq \mathrm{Gal}(KL/\mathbb{Q})$ (cf. Figure 9.4), and $\mathfrak{P}'' = \mathfrak{P}' \cap K'$, then $\mathfrak{P}''$ is unramified over p. Further,

$$K' \cap L = \mathbb{Q},$$

because if we put $k = K' \cap L$, then p is both unramified and totally ramified in k, which can happen only if $k = \mathbb{Q}$.

Now, because K' is the fixed field of the subgroup T' of order l^a, we have $[KL : K'] = l^a$. Therefore,

$$[KL : \mathbb{Q}] = [KL : K'][K' : \mathbb{Q}] = l^a[K' : \mathbb{Q}].$$

Also, because $K' \cap L = \mathbb{Q}$,

$$[K'L : \mathbb{Q}] = [K' : \mathbb{Q}][L : \mathbb{Q}] = [K' : \mathbb{Q}]l^a.$$

Hence $K'L = KL$.

Finally, $p = p_r$ is unramified in K', and if a prime $q \neq p_1, \ldots, p_{r-1}, l$, then q is unramified in K'. Further, $K'/\mathbb{Q}$ is cyclic of degree l^s. Thus, by induction hypothesis, K' is cyclotomic, hence so is $K'L = KL \supseteq K$. $\square$

Theorem 9.11 (Kronecker-Weber). *If K is a finite Abelian extension of $\mathbb{Q}$, then K is cyclotomic.*

Proof. By the Fundamental Theorem of Abelian groups, $G = \mathrm{Gal}(K/\mathbb{Q})$ is a direct product of cyclic groups of prime power orders, say

$$G = C_1 \times \cdots \times C_r. \tag{9.9}$$

Let G_j be the subgroup of G, obtained from (9.9) on replacing the jth factor C_j by $\{1\}$. We put $K_j = K^{G_j}$, the fixed field of G_j. Then $\mathrm{Gal}(K_j/\mathbb{Q}) \cong C_j$, so each K_j is cyclotomic. But K is the compositum of $K_1, \ldots, K_r$. Hence K is also cyclotomic. $\qquad\square$

10

Passage to Algebraic Geometry

In this chapter, we indicate how the subject treated in the previous chapters leads naturally to the study of arithmetic algebraic geometry.

Dedekind's generalization (in 1870) of numbers and primes in $\mathbb{Z}$ to ideals and nonzero prime ideals in the ring $\mathcal{O}_K$ of integers in a number field K is a crowning achievement in number theory. Dedekind was a student of Gauss. Riemann, another student of Gauss, made (in 1859) probably the most famous conjecture in all of mathematics, called the Riemann Hypothesis. He assumed it in order to count the number of primes $p \leq x$, conjectured by Legendre and Gauss. The zeta function he used for this purpose had already been studied by Euler as a function

$$\zeta(\sigma) = \sum_{n=1}^{\infty} \frac{1}{n^{\sigma}} \tag{10.1}$$

of a real variable σ.

The series in (10.1) converges for $\sigma > 1$. Riemann studied it as a function of a complex variable $s = \sigma + it$, namely

$$\zeta(s) = \sum_{n=1}^{\infty} \frac{1}{n^{s}} . \tag{10.2}$$

The series on the right of (10.2) converges for $\mathrm{Re}(s) > 1$. He showed that the function defined by (10.2) for $\mathrm{Re}(x) > 1$ extends by a functional equation to an analytic function on the whole complex plane $\mathbb{C}$ except for a simple pole at $s = 1$. The function $\zeta(s)$ is called the *Riemann zeta function*. The *Riemann Hypothesis* asserts that all its nonreal zeros lie on the line $\sigma = \frac{1}{2}$.

The series on the right of (10.2) has the *Euler product*

$$\sum_{n=1}^{\infty} \frac{1}{n^{s}} = \prod_{p} \left(1 - \frac{1}{p^{s}}\right)^{-1} . \tag{10.3}$$

The Dedekind zeta function $\zeta_K(s)$ of a number field K is defined analogously by

$$\zeta_K(s) = \sum_{\mathfrak{a}} \frac{1}{N(\mathfrak{a})^{s}} ,$$

the sum taken over all nonzero ideals $\mathfrak{a}$ of $\mathcal{O}_K$. This too has the Euler product

$$\zeta_K(s) = \prod_{\mathfrak{p}} \left(1 - \frac{1}{N(\mathfrak{p})^s}\right)^{-1}, \tag{10.4}$$

the product taken over all prime ideals $\neq (0)$ of $\mathcal{O}_K$. Clearly, it is a generalization of the Riemann zeta function and counts the number of prime ideals $\mathfrak{p}$ in $\mathcal{O}_K$ with $N(\mathfrak{p}) \leq x$, for x arbitrarily large. The Riemann Hypothesis can be extended to $\zeta_K(s)$, for this count, to assert that its nonreal zeros have $\mathrm{Re}(s) = \frac{1}{2}$.

During the earlier part of the 20th century it was realized that every zeta function counts something of interest to number theorists, if defined properly. For example, if we have a curve C defined over a finite field $\mathbb{F}_q$ of q elements by an irreducible polynomial equation

$$f(x, y) = 0, \tag{10.5}$$

there are only finitely many points on C with coordinates in the algebraic closure of $\mathbb{F}_q$ of a "bounded size," which can be counted by the zeta function $\zeta_C(s)$ of the curve C.

For a clue on how to define $\zeta_C(s)$, we write (10.5) as

$$a_n(x)y^n + \cdots + a_1(x)y + a_0(x) = 0 \tag{10.6}$$

with $a_j(x)$ in the field $\mathbb{F}_q(x)$ of rational functions over $\mathbb{F}_q$, to get an extension K of $\mathbb{F}_q(x)$ of finite degree, analogous to that of $\mathbb{Q}$.

Once it was realized by Ostrowski, Artin and others that the primes in $\mathbb{Z}$, or more generally the prime ideals $\mathfrak{p}$ in the ring $\mathcal{O}_K$ of integers of a number field K are in a one-to-one correspondence with (inequivalent) valuations on K, the Dedekind zeta function was defined entirely in terms of the valuations on K. The valuation on the function field $\mathbb{F}_q(x, y) = $ quotient field of $\mathbb{F}_q[x, y]/(f(x, y))$ of the curve C defined by (10.5) are given by the points on C. Thus to define the zeta function $\zeta_C(s)$ of the curve C and to propose the Riemann Hypothesis for $\zeta_C(s)$, one appeals to the points on C. Later in the chapter, we will prove the Riemann Hypothesis for curves of genus 1 over $\mathbb{F}_q$ and show that it is intimately related to counting points on these curves.

Surprisingly, the Riemann Hypothesis for curves over finite fields has been shown to be true, by Hasse for curves of genus 1 and by Weil in general. Moreover, its generalization to varieties over finite fields (solutions of one or more polynomial equations in $n(n \geq 2)$ variables over finite fields) conjectured by Weil, was proved by Deligne in 1974 for which he was awarded the Fields Medal. However, the original Riemann Hypothesis remains unproved.

In view of the above discussion, in order to study the number fields and function fields of curves over finite fields as a single subject, by a *global field*

K it is meant that K is either a number field or the function field of a curve over a finite field.

10.1 Valuations

Suppose K is a field (not necessarily a global field). We denote the nonzero elements of K by $K^\times$. A *discrete valuation* on K is a map $v : K^\times \to \mathbb{Z}$, such that for all x, y in $K^\times$,

1. $v(xy) = v(x) + v(y)$
2. $v(x + y) \geq \min(v(x), v(y))$.

As a matter of convenience, we extend a valuation to a map $v : K \to \mathbb{Z} \cup \{\infty\}$ by setting $v(0) = \infty$. Throughout this chapter we will exclude the trivial valuation given by the zero map.

Two valuations on K are *equivalent valuations* if they can be scaled to give the same valuation. Among all the equivalent valuations, there is a unique one which is surjective. We use this *normalized valuation* as a representative of all valuations equivalent to it.

Example 10.1. *The p-adic valuation*

Let $K = \mathbb{Q}$ and $p = 2, 3, 5, \ldots$ a fixed prime. Any $x \neq 0$ in $\mathbb{Q}$ has a unique representation
$$x = p^{v_p(x)} \frac{a}{b},$$
where $\frac{a}{b}$ ($b > 0$) is in the lowest form, p and ab have no nontrivial common factor and v_p is in $\mathbb{Z}$. It is easy to check that v_p is a valuation on $\mathbb{Q}$, called the *p-adic valuation*.

The p-adic valuations on $\mathbb{Q}$ can be extended in an obvious way to $\mathfrak{p}$-*adic valuations on a number field* K, $\mathfrak{p}$ being a nonzero prime ideal of $\mathcal{O}_K$. For a nonzero x in K write the principal fractional ideal $(x) = x\,\mathcal{O}_K$ generated by x as
$$(x) = \mathfrak{p}^{v_\mathfrak{p}(x)} \frac{\mathfrak{a}}{\mathfrak{b}}$$
with $\mathfrak{p}$ not appearing in the unique factorization of $\mathfrak{a}$ and $\mathfrak{b}$ as products of powers of distinct prime ideals of $\mathcal{O}_K$. It is a trivial consequence of Theorem 3.69 (independence of valuations) that different prime ideals give inequivalent $\mathfrak{p}$-adic valuations.

Discrete Valuation Rings

For a discrete valuation v on a field K, let

$$\mathcal{O}_v = \{x \in K \mid v(x) \geq 0\}$$

and

$$\mathfrak{p}_v = \{x \in K \mid v(x) > 0\}.$$

Then $\mathcal{O}_v$ is a ring, called the *discrete valuation ring* of v and $\mathfrak{p}_v$ is a prime ideal of $\mathcal{O}_v$. If the quotient $\mathcal{O}_v/\mathfrak{p}_v$ is finite, we call the cardinality $|\mathcal{O}_\mathfrak{p}/\mathfrak{p}_v|$ the *norm of the valuation* v and denote it by N_v.

In some situations, it is preferable to work with the multiplicative version of the discrete valuations on K, called the *absolute values* on K.

Definition 10.2. An *Archimedean absolute value* on a field K is a map $|\,| : K \to [0, 1)$ such that

1. $|x| = 0 \Leftrightarrow x = 0$,

2. $|xy| = |x|\,|y|$ and

3. *Triangle Inequality* holds: $|x + y| \leq |x| + |y|$.

Clearly the restriction of $|\,|$ to $K^\times$, the group of nonzero elements of K, is a group homomorphism from $K^\times$ to the multiplicative group $\mathbb{R}^+$ of positive reals.

Example 10.3. For $z = x + iy$ in $\mathbb{C}$, $|z| = \sqrt{x^2 + y^2}$ is an Archimedean absolute value on $\mathbb{C}$. If K is a subfield of $\mathbb{C}$, it inherits this absolute value from $\mathbb{C}$. The theorem of Gelfand and Tornheim [1] asserts that it is essentially the only Archimedean absolute values on K.

Definition 10.4. A *non-Archimedean absolute value* on a field K is a map $|\,| : K \to [0, 1)$ with

1. $|x| = 0 \Leftrightarrow x = 0$,

2. $|xy| = |x|\,|y|$ and

3. a stronger inequality $|x + y| \leq \max(|x|, |y|)$ holds.

By an absolute value we shall mean a non-Archimedean absolute value, unless to the contrary is clear from the context. If $|x| = 1$ for all $x \neq 0$, the absolute value is *trivial*. It will be excluded from our discussion. Two absolute values $|\,|_1$ and $|\,|_2$ are *equivalent absolute values* if for a constant $c > 0$, $|x|_2 = |x|_1^c$ for all x in K. We can define a distance $\mathrm{dist}(x, y)$ between points x, y and K by $\mathrm{dist}(x, y) = |x - y|$. The equivalent valuations thus define the same topology on K. In fact, two absolute values are equivalent if and only if they induce the same topology. For the fields that concern us, there is a natural choice for the normalized absolute values on K.

Example 10.5. 1. For $K = \mathbb{Q}$ the *p-adic absolute value* is given by

$$|x|_p = p^{-v_p(x)}$$

for $x \neq 0$.

2. If K is a number field, $\mathfrak{p}$ a nonzero prime ideal of $\mathcal{O}_K$, the $\mathfrak{p}$-*adic absolute value* on K is defined by

$$|x|_\mathfrak{p} = N(\mathfrak{p})^{-v_\mathfrak{p}(x)}.$$

In particular, if $K = \mathbb{Q}$, $N(p\mathbb{Z}) = p$ and thus $\mathfrak{p}$-adic absolute value on K is a generalization of the p-adic absolute value on $\mathbb{Q}$.

The following result complements the theorem of Gelfand and Tornheim.

Theorem 10.6 (Ostrowski). *The only (inequivalent) non-Archimedean absolute values on a number field K are the $\mathfrak{p}$-adic absolute values on K.*

Proof. We prove it for $K = \mathbb{Q}$ and leave it as an exercise to extend it to a number field K. It is immediate by 2) of Definition 10.4 that $|\pm 1| - 1$. If $n > 0$ is an integer, by 3), $|n| = |1 + \cdots + 1| \leq 1$. Because $|\ |$ is nontrivial, $|n| < 1$ for some $n \neq 0$. In fact, $|p| < 1$ for a prime p, for otherwise by 2), $|\ |$ would be trivial on $\mathbb{Q}$.

We now show that if $q \neq p$ is another prime, then $|q| = 1$. Suppose $|q| < 1$. Choose integers a and b such that $1 = ap + bq$. Then

$$1 = |1| = |ap + bq| \leq \max(|p|, |q|) < 1.$$

This is a contradiction.

Finally, put $|p| = c > 0$. Then $|x| = c^{v_p(x)}$, hence $|\ |$ is equivalent to $|\ |_p$. $\square$

Up to equivalence, there is only one Archimedean absolute value on a number field $\mathbb{Q}$, the one inherited from the absolute value on $\mathbb{C}$ (Gelfand-Tornheim). It is convenient to denote it by $|x|_\infty$ and regard it to be the one that corresponds to the *infinite prime* $p = \infty$.

Theorem 10.7 (Product Formula). *For $x \neq 0$ in $\mathbb{Q}$,*

$$\prod_p |x|_p = 1,$$

where $p = \infty, 2, 3, 5, \ldots$.

Proof. The proof follows at once from the unique factorization in $\mathbb{Q}$ and the definition of $|\ |_p$ for $p = \infty, 2, 3, 5, \ldots$. $\qquad\qquad\square$

Remark 10.8. The product formula extends to number fields in a fairly obvious way.

Reformulation of Dedekind Zeta Function

Let V_K be the set of all (inequivalent) non-Archimedean absolute values on a field K. For each v in V_K, we define two subsets of K by

$$\mathcal{O}_v = \{v \in K \mid |x|_v \leq 1\} \cup \{0\}$$

and

$$\mathfrak{p}_v = \{x \in K \mid |x|_v < 1\} \cup \{0\}.$$

It is easily seen that $\mathcal{O}_v$ is a ring and $\mathfrak{p}_v$ a prime ideal of $\mathcal{O}_v$. Thus we can make the quotient ring $\mathcal{O}_v/\mathfrak{p}_v$. If its cardinality $|\mathcal{O}_v/\mathfrak{p}_v|$ is finite, we let $N_v = |\mathcal{O}_v/\mathfrak{p}_v|$, the *norm* of v. If K is a global field, this is always the case. In particular, if K is a number field, we identify v with the corresponding prime ideal $\mathfrak{p}$ of K. Then $N_v = |\mathcal{O}/\mathfrak{p}| = N(\mathfrak{p})$, the norm of $\mathfrak{p}$. So we can reformulate the Dedekind zeta function as

$$\zeta_K(s) = \prod_{v \in V_K} \left(1 - \frac{1}{N_v^s}\right)^{-1}. \tag{10.7}$$

10.2 Zeta Functions of Curves over Finite Fields

We begin with a curve defined by an irreducible equation

$$f(x, y) = 0 \tag{10.8}$$

over any given field k. It is *smooth* or *non-singular* if the partial derivatives $\frac{\partial f}{\partial x}$ and $\frac{\partial f}{\partial y}$ do not vanish simultaneously at any point on C, including the points at infinity on it. The *genus* g of such a curve is defined by

$$g = \frac{(n-1)(n-2)}{2}, \tag{10.9}$$

where $n = \deg f(x, y)$.

We assume all our curves to be smooth. Then the genus is 1 if and only if $f(x, y)$ is a cubic. For the sake of simplicity in our exposition, we assume that in case k is a finite field, it is not an extension of $\mathbb{F}_2$ or $\mathbb{F}_3$. Then by a suitable substitution for the variables, a curve of genus 1 having a point with coordinates in k is defined by a *Weierstrass equation*

$$y^2 = x^3 + ax + b \quad (a, b \in k) \tag{10.10}$$

with the *discriminant* $\Delta = -16(4a^3 + 27b^2) \neq 0$.

For the so-called *geometric case* of global fields, one takes $f(x, y)$ in (10.8) with coefficients in a finite field $\mathbb{F}_q$ of a prime power $q = p^a (a \geq 1)$ elements. Then this global field is the function field of C, which by definition is the quotient field of the quotient ring $\mathbb{F}_q[x, y]/(f(x, y))$. It is an integral domain for $f(x, y)$ is irreducible. If we have rational functions $\frac{g_1(x,y)}{h_1(x,y)}$ and $\frac{g_2(x,y)}{h_2(x,y)}$ where numerators and denominators agree modulo $f(x, y)$, they define the same function on C. Therefore, for each point P on C, with coordinates in the algebraic closure $\overline{\mathbb{F}}_q$ of $\mathbb{F}_q$, we can define a discrete valuation v_P on K as follows:

$$v_P(f) = \begin{cases} \text{the order of zero of } f \text{ at } P \\ - \text{ the order of pole of } f \text{ at } P \\ 0 \text{ if } P \text{ is neither a zero nor a pole of } f. \end{cases}$$

Since a rational function has only finitely many zeros and poles, $v_P(f) = 0$ for all except finitely many P.

It is a standard fact from algebraic geometry that for any f in K, the number of its zeros is equal to the number of its poles, counted with multiplicities, i.e.

$$\sum_{P \in C} v_P(f) = 0 \tag{10.11}$$

but not as stated here. Recall that the multiplicative version of (10.11), namely Theorem 10.7, is not valid until the absolute value $| \,|_p$ for the infinite prime $p = \infty$ is taken into account. Similarly, for (10.11) to hold, the summation above should include the points at infinity on C.

To explain what the points at infinity on C are, one homogenizes its equation (10.8) by putting $x = \frac{X}{Z}$, $y = \frac{Y}{Z}$ in it and clears the denominators to get the homogeneous polynomial equation

$$F(X, Y, Z) = Z^{\deg f} f\left(\frac{X}{Z}, \frac{Y}{Z}\right) = 0 \tag{10.12}$$

and require that $(X, Y, Z) \neq (0, 0, 0)$. Then (X, Y, Z) is a solution $\Leftrightarrow$ for each $c \neq 0$, $c(X, Y, Z)$ is a solution of (10.12). All such solutions are regarded as one point on C. The solutions to (10.8) are the points with $Z = 1$ and the *points at infinity* on C are those with $Z = 0$. The curve C is a *complete curve* when the points at infinity on it are added to it.

Example 10.9. Let E be the *elliptic curve* (by definition a curve of genus 1, given by its Weierstrass equation (10.10)).

By substituting $x = \frac{X}{Z}$, $y = \frac{Y}{Z}$ in (10.10) and clearing the denominators from the resulting equation, we get the homogeneous equation

$$Y^2 Z = X^3 + aXZ^2 + bZ^3$$

of degree 3. If $Z = 1$, we get

$$Y^2 = X^3 + aX + b \qquad (10.13)$$

which is the same equation as (10.10). Thus we recover all the points on E defined by (10.10). If $Z = 0$, then $X = 0$. But since $XYZ \neq 0$, $Y \neq 0$, which may be taken to be 1. Hence E has only one point at infinity with homogeneous coordinates $(0 : 1 : 0)$. Moreover, $(0 : 1 : 0) = (0 : \pm t : 0)$ for t however large we want, so $(0 : 1 : 0) = (0 : \pm\infty : 0)$ is the point at infinity on E. Intuitively, one may think of it as two points at infinity on either end of the y-axis.

Singularities

A point P on a curve C is *singular point* if all the partial derivatives

$$\frac{\partial F}{\partial X} = \frac{\partial F}{\partial Y} = \frac{\partial F}{\partial Z}$$

vanish at P, i.e. we cannot write the equation of the tangent line to C at P. The curve C is *non-singular* or *smooth* if it has a tangent at every point on it, including at the points at infinity on it.

Exercise 10.10. Show that the cubic defined by (10.10) is smooth if and only if $x^3 + ax + b$ has distinct roots. [If a, b in a field k, assume that k is not an extension of $\mathbb{F}_2$ or $\mathbb{F}_3$.]

There is another subtlety that needs to be addressed. If we allow, which we will, the points on C to have coordinates in every finite field containing $\mathbb{F}_q$ (the field of definition of C), different points on C can define the same v_P.

Example 10.11. To be concrete let us take $q = 3$ and $f(x, y) = y$. Then $\mathbb{F}_3[x, y]/(f(x, y)) \cong \mathbb{F}_3[x]$ and the function field of the curve C defined by

$$f(x, y) = 0$$

is $K = \mathbb{F}_3(x)$. Therefore, C is the x-axis. For the sake of completeness we include the single point on x-axis at infinity. Thus C is the projective line $\mathbb{P}^1$ over $\mathbb{F}_3$.

Since -1 is not a square in $\mathbb{F}_3$, the function $f(x) = x^2 + 1$ has no zero in $\mathbb{F}_3$. However, it factors as $f(x) = (x + i)(x - i)$ in $\mathbb{F}_3(i)$, $i = \sqrt{-1}$. There are thus two points $P = (x) = (\pm i)$ on C with coordinates in the quadratic extension $\mathbb{F}_3(i)$ of $\mathbb{F}_3$. For each of these points $v_P(f) = 1$.   ∎

In general, the discrete valuations on the function field K of a curve C are given by points on the (complete) curve C, but if a point P on C has coordinates in $\mathbb{F}_{q^d}$ but not in any smaller field, then there are d points P on C with the same valuation v_P. For such points, $N_{v_P} = q^d$. By definition, $d = \deg(P)$. Now we can say that (10.11) is valid if C is complete and the points P have coordinates in the algebraic closure $\overline{\mathbb{F}}_q$ of $\mathbb{F}_q$.

Zeta Functions of Varieties over Finite Fields

Let K be an algebraically closed field. A *closed set* in $K^n = \underbrace{K \times \cdots \times K}_{n \text{ times}}$ is the set X of solutions in K^n to a finite number of polynomial equations

$$f_j(x_1, \ldots, x_n) = 0, \quad j = 1, \ldots, m \tag{10.14}$$

with coefficients in K. If these coefficients are in a subfield k of K, we say X is *defined over k*. A closed set V is a *variety* if it is irreducible, i.e. V is not a union $V = X_1 \cup X_2$ of proper closed subsets X_j of V. A variety V is *smooth* if at every point, it has a well-defined tangent plane. The *dimension* $\dim(V)$ of V is the number of free variables in (10.14) and it is *projective* if its points have coordinated in the projective space $\mathbb{P}^n(K)$.

Now let $k = \mathbb{F}_q$ and $K = \overline{\mathbb{F}}_q$, the algebraic closure of $\mathbb{F}_q$. Every element of $\overline{\mathbb{F}}_q$ is an element of a finite extension $\mathbb{F}_{q^r}$ of $\mathbb{F}_q$ for some $r \geq 1$. Let $N_r(V)$ be the number of points of V with coordinates in $\mathbb{F}_{q^r}$. The zeta function of V has several equivalent formulations. The one we have been aiming for is

$$\zeta_V(s) = \prod_{v \in V_K} \left(1 - \frac{1}{N_v^s}\right)^{-1}, \tag{10.15}$$

V_K being the set of (inequivalent) valuations on K, the function field of V. It is easily seen that

$$\zeta_V(s) = \exp\left(\sum_{r=1}^{\infty} \frac{N_r(V)}{r} (q^{-s})^r\right). \tag{10.16}$$

Following Weil, one puts $t = q^{-s}$ and writes (10.16) as

$$Z_V(t) = \exp\left(\sum_{r=1}^{\infty} \frac{N_r(V)}{r} t^r\right).$$

In 1949, Weil [38] made the following conjectures about $Z_V(t)$ after proving it when $\dim(V) = 1$, i.e. for curves.

Weil Conjectures

Let V be a smooth projective variety of dimension d over $\mathbb{F}_q$. Then

1. $Z_V(t)$ is a rational function of t of the form

$$Z_V(t) = \frac{P_1(t)P_3(t)\cdots P_{2d-1}(t)}{P_0(t)P_2(t)\cdots P_{2d}(t)},$$

where $P_0(t) = 1 - t$, $P_{2d}(t) = 1 - q^{dt}$ and each $P_i(t)$ is a polynomial over $\mathbb{Z}$ factoring over $\mathbb{C}$ as $P_i(t) = \prod_j (1 - \alpha_{ij}t)$.

2. If E is the Euler characteristic of V, there is the functional equation

$$Z_V\left(\frac{1}{q^d t}\right) \pm q^{dE/2}t^E Z_V(t).$$

3. *Riemann Hypothesis.* If $Z_V(t) = 0$, then $|\alpha_{ij}| = q^{i/2} \Rightarrow \mathrm{Re}(s) = \frac{1}{2}$.

For elliptic curves (curves of genus 1) over finite fields, this was conjectured by E. Artin [1] in his thesis (1924) and proved in 1936 by Hasse. The rationality of the zeta functions of curves in general was proved in 1931 by F.K. Schmidt [34] whereas for varieties of higher dimensions, it was shown to be true by B. Dwork [14] in 1960. Finally, the last and the most difficult part of the Weil conjectures was proved by Deligne [12] in 1974, for which he was awarded the Fields Medal. The scheme theoretic algebraic geometry [20] was developed primarily for this purpose.

We will discuss the Weil conjecture for the simplest nontrivial case (elliptic curves over finite fields) and prove the Riemann Hypothesis in this case. For a curve of higher genus, see [4], [35] and [36]. For Deligne's proof of the Weil conjecture, see [16].

Projective Space $\mathbb{P}^d$

Let V be the projective space $\mathbb{P}^d(\overline{\mathbb{F}}_q)$ of dimension d with $\overline{\mathbb{F}}_q$, the algebraic closure of $\mathbb{F}_q$. As an example, we compute its zeta function. But before doing so, let us be clear what it is.

Let K be any field. The *projective space* $\mathbb{P}^d(K)$ consists of nonzero points $(x_0, x_1, \ldots, x_d)$ of K^{d+1} with two such points $\boldsymbol{x} = (x_0, x_1, \ldots, x_d)$, $\boldsymbol{y} = (y_0, y_1, \ldots, y_d)$ representing the same point of $\mathbb{P}^d(K)$ if $\boldsymbol{y} = c\boldsymbol{x}$ for a nonzero c in K. In other words, $\boldsymbol{x}$ and $\boldsymbol{y}$ represent the same point P of $\mathbb{P}^d(K)$ if their coordinates are proportional. For this reason, we write $P = (x_0 : x_1 : \cdots : x_d)$. If one such representative satisfies a homogeneous equation in $d+1$ variables, so do all other representatives of P. Thus it makes sense to talk about the solutions to a homogeneous polynomial equation in $\mathbb{P}^d(K)$. The set of solutions in $P^d(K)$ to a finite number of homogeneous polynomial equations in $d+1$ variable is called a *projective variety*. In particular, $\mathbb{P}^d(K)$ is a projective variety defined by the zero polynomial (one with all its coefficients $= 0$). If the coefficients of these polynomials are in a subfield k of K, we say the variety is *defined over* k. For a field L such that $k \subseteq L \subseteq K$, the set of points in $\mathbb{P}^d(L)$ of a variety V defined over k is denoted by $V(L)$.

Now let $k = \mathbb{F}_q$, $K = \overline{\mathbb{F}}_q$ and $V = \mathbb{P}^d$. Then

$$N_r(V) = |V(\mathbb{F}_{q^r})| = \frac{q^{r(d+1)} - 1}{q^r - 1} = \sum_{j=0}^{d} q^{jr}.$$

Hence

$$\log Z_V(t) = \sum_{r=1}^{\infty} \left(\sum_{j=0}^{d} q^{jr} \right) \frac{t^r}{r}$$

$$= \sum_{j=0}^{d} \left(\sum_{r=1}^{\infty} \frac{(q^j t)^r}{r} \right) = -\sum_{j=0}^{d} \log(1 - q^j t)$$

$$= \log \prod_{j=0}^{d} (1 - q^j t)^{-1}.$$

Applying the exp map to each side we get

$$Z_V(t) = \frac{1}{(1 - t)(1 - qt) \dots (1 - q^d t)}.$$

10.3   Riemann Hypothesis for Elliptic Curves over Finite Fields

We denote by E the elliptic curve defined by its Weierstrass equation

$$y^2 = x^3 + ax + b$$

i.e. with a, b in $\mathbb{F}_q$, q a power of a prime p.

As said, its zeta function $Z_E(t)$ has something to do with counting points on E. To see what to expect, suppose the values of $x^3 + ax + b$ are evenly distributed as x varies over $\mathbb{F}_q$. When $x^3 + ax + b = 0$, we get one point of E. Because q is odd, half of the $q - 1$ nonzero values of $x^3 + ax + b$ are expected to be nonsquare giving no point of E. For the other half, we have

$(\pm y)^2 = x^3 + ax + b$ for some y in $\mathbb{F}_q$, giving two points of E for each of the $\frac{q-1}{2}$ x in $\mathbb{F}_q$. Thus the expected number of points $N_q = |E(\mathbb{F}_q)| = 1 + 2 \cdot \frac{q-1}{2} = q$. We define the integer

$$a_q = a_q(E) = q - N_q \tag{10.17}$$

to be the deviation of N_q from the expected number q of points on $E(\mathbb{F}_q)$. It then turns out that (cf. [37, Prop. 12.1])

$$Z_E(t) = \frac{1 - a_q(E)t + qt^2}{(1 - t)(1 - qt)} . \tag{10.18}$$

In 1936, H. Hasse proved the estimate

$$|a_q(E)| \leq 2\sqrt{q} \tag{10.19}$$

for $a_q(E)$, which is called the Riemann Hypothesis for elliptic curves over finite fields for the following reason.

Recall the convention $t = q^{-s}$. Thus if $Z_E(t) = 0$, then q^s is a root of the polynomial

$$f(u) = u^2 - a_q u + q.$$

The inequality (10.19) holds if and only if the discriminant $a_q^2 - 4q$ of $f(u)$ is ≤ 0, which is true if and only if the two roots u_1, u_2 of $f(u)$ are either real and equal, or are a pair of complex conjugates. Since the constant term q of $f(u)$ is the product $u_1 u_2$, (10.19) holds if and only if both roots of $f(u)$ have absolute value $\sqrt{q}$, if and only if for all s with $Z_E(q^{-s}) = 0$, $|q^s| = \sqrt{q}$. This implies that $\mathrm{Re}(s) = 1/2$.

Broadly speaking, it is the geometric interpretations that have led to the proof of the Riemann Hypothesis, in algebraic geometry, while the case of Riemann's original zeta function remains so intractable.

An Elementary Proof of Hasse's Theorem

We now prove the Hasse theorem (inequality (10.19)), equivalently the Riemann Hypothesis for elliptic curves over finite fields. The proof is essentially that of Manin [26], which in itself is based on the original one by Hasse. To begin with let us assume that k is any field that does not contain $\mathbb{F}_2$ or $\mathbb{F}_3$ as a subfield. Then, an elliptic curve E over k is defined by (10.10):

$$y^2 = x^3 + ax + b \quad (a, b \in k)$$

with $4a^3 + 27b^2 \neq 0$.

If K is any field containing k, then the set $E(K)$ consisting of points on it with coordinates in K together with its point O at infinity forms an Abelian group. For $k = \mathbb{Q}$ and $K = \mathbb{R}$, assuming $x^3 + ax + b$ has only one real root, it looks as in Figure 10.1.

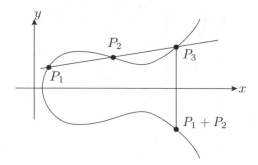

FIGURE 10.1: Adding points on elliptic curves.

As observed earlier, the point O at infinity is on each end of every vertical line. The sum of two points P_1, P_2 is the reflection in the x-axis of the third point P_3 of intersection of the line through P_1 and P_2 (tangent to (10.10) at P if $P_1 = P_2 = P$) with the cubic (10.10). One then checks that O is the zero of the group, and that the inverse of a point (x,y) is given simply by $(x,-y)$.

Twists

To prove the inequality (10.19), we shall be working with another elliptic curve closely related to E. It is defined over the function field $K = \mathbb{F}_q(t)$ by

$$\lambda y^2 = x^3 + ax + b, \tag{10.20}$$

where $\lambda = \lambda(t) = t^3 + at + b$. The elliptic curve E_λ given by equation (10.20) is a *twist* of E.

If $x(P)$ denotes the x-coordinate of a point P, we compute $x(P_1 + P_2)$ for P_1, P_2 in $E_\lambda(K) = \{(x,y) \in K^2 \mid \lambda y^2 = x^3 + ax + b\} \cup \{O\}$. This formula for $x(P_1 + P_2)$ in terms of $x(P_1)$ and $x(P_2)$ plays a dominant role in the proof of inequality (10.19). We leave aside certain cases (such as $x(P_1) = x(P_2)$; P_1 or $P_2 = O$) that we do not need for our proof.

Suppose $P_j = (X_j, Y_j) \in E_\lambda(K)$ for $j = 1, 2$. To compute $x(P_1 + P_2)$ we write the equation of the line through P_1 and P_2, which is

$$y = \left(\frac{Y_1 - Y_2}{X_1 - X_2} \right) x + \ell \tag{10.21}$$

To find the x-coordinate X_3 of the third point P_3 of intersection of this line with the cubic (10.20), we substitute for y from (10.21) in (10.20) to get

$$x^3 - \lambda \left(\frac{Y_1 - Y_2}{X_1 - X_2} \right)^2 x^2 + \cdots = 0. \tag{10.22}$$

Since X_1, X_2, X_3 are the three solutions of (10.22), the left side of (10.22) is

$$(x - X_1)(x - X_2)(x - X_3)$$
$$= x^3 - (X_1 + X_2 + X_3)x^2 + \cdots . \tag{10.23}$$

Comparing the coefficient of x^2 in (10.22) and (10.23), we get

$$x(P_1 + P_2) = X_3 = \lambda \left(\frac{Y_1 - Y_2}{X_1 - X_2} \right)^2 - (X_1 + X_2). \tag{10.24}$$

Frobenius map

A crucial ingredient in the proof of inequality (10.19) is the *Frobenius map* Φ and its elementary properties. For a fixed q, let K be any field containing $\mathbb{F}_q$ as a subfield. We define $\Phi = \Phi_q : K \to K$ as the function given by $\Phi(X) = X^q$.

We summarize the properties of the Frobenius map.

Theorem 10.12. *The Frobenius map $\Phi(X) = X^q$ has the following properties:*

> *i)* $(XY)^q = X^q Y^q$.
>
> *ii)* $(X + Y)^q = X^q + Y^q$.
>
> *iii)* $\mathbb{F}_q = \{\alpha \in K \mid \Phi(\alpha) = \alpha\}$.
>
> *iv)* *For $\phi(t)$ in $\mathbb{F}_q(t)$, $\Phi(\phi(t)) = \phi(t^q)$.*

Although it is not used directly in this proof of the Hasse inequality, it is worth noting that iii) above implies that $E(\mathbb{F}_q)$ consists precisely of points fixed by Φ. Other proofs of the Hasse inequality use this fact directly.

Proof. i) is trivial.

ii) We use induction on $r = \log_p q$. If $r = 1$, $q = p$ and

$$(X + Y)^p = \sum_{j=0}^{p} \binom{p}{j} X^j Y^{p-j}.$$

For $0 < j < p$, the binomial coefficient $\binom{p}{j}$ satisfies

$$\binom{p}{j} = \frac{p!}{j!(p - j)!} = p \cdot m$$

for some positive integer m, because nothing in the denominator can cancel p in the numerator and $\binom{p}{j}$ is a whole number. Since $p\alpha = 0$ for all α in K, ii) follows. For $r > 1$, by the induction hypothesis

$$(X + Y)^q = ((X + Y)^{p^{r-1}})^p$$
$$= (X^{p^{r-1}} + Y^{p^{r-1}})^p$$
$$= X^q + Y^q.$$

iii) The set $\mathbb{F}_q^\times$ of nonzero elements of $\mathbb{F}_q$ is a multiplicative group of order $q - 1$. Therefore, by elementary group theory, $\alpha^{q-1} = 1$ for all α in $\mathbb{F}_q^\times$. In other words, each of the q elements of $\mathbb{F}_q$ is a root of the polynomial $t^q - t = t(t^{q-1} - 1)$ of degree q. Since a polynomial of degree q cannot have more than q roots, $\mathbb{F}_q$ consists precisely of the elements of K which are roots of $t^q - t$. This proves iii).

iv) follows at once from i), ii) and iii). $\qquad\square$

Counting points on elliptic curves

We now use the properties of the Frobenius map Φ_q and the group law on the elliptic curve $E_\lambda(K)$ to count the number of solutions of the equation $y^2 = x^3 + ax + b$ $(a, b \in \mathbb{F}_q,\ q = p^r,\ 4a^3 + 27b^2 \neq 0)$ with x, y in $\mathbb{F}_q$.

Clearly $(t, 1)$ and its negative $-(t, 1) = (t, -1)$ are in $E_\lambda(K)$. Using the properties of Φ_q, it is also clear that the point

$$P_0 = (t^q, (t^3 + at + b)^{(q-1)/2})$$

is in $E_\lambda(K)$.

We define a degree function d, which we will show to be a quadratic polynomial with nonreal roots. Its discriminant plays a central role in the proof of inequality (10.19). For $n \in \mathbb{Z}$, let

$$P_n = P_0 + n(t, 1),$$

the addition being the one on $E_\lambda(K)$. Define $d : \mathbb{Z} \to \{0, 1, 2, \ldots\}$ by

$$d(n) = d_n = \begin{cases} 0, & \text{if } P_n = O; \\ \deg(\text{num}(x(P_n))), & \text{otherwise.} \end{cases}$$

Here $\text{num}(X)$ is the numerator of a rational function $X \in \mathbb{F}_q(t)$, taken in the lowest form. The values of this degree function on three consecutive integers satisfy (for a proof, see [10]) the following identity:

Basic Identity

$$d_{n-1} + d_{n+1} = 2d_n + 2. \tag{10.25}$$

The crux of the proof of (10.19) is the following theorem relating the degree function to the number N_q of solutions of $y^2 = x^3 + ax + b$ with x, y in $\mathbb{F}_q$.

Theorem 10.13.
$$d_{-1} - d_0 - 1 = N_q - q. \tag{10.26}$$

Proof. Let $X_n = x(P_n)$. Since $P_0 \neq (t, 1)$, we have $P_{-1} \neq O$, so $d_{-1} = \deg(\operatorname{num}(X_{-1}))$. We therefore compute X_{-1} and look at the degree of its numerator when it is in the lowest form. By (10.24),

$$X_{-1} = \frac{(t^3 + at + b)[(t^3 + at + b)^{(q-1)/2} + 1]^2}{(t^q - t)^2} - (t^q + t) \tag{10.27}$$

$$= \frac{t^{2q+1} + \text{ lower terms}}{(t^q - t)^2},$$

where the last expression is obtained by putting the previous one over the common denominator $(t^q - t)^2$ and using property iv) of the Frobenius map. We must cancel any common factors in the last expression. Since the term $t^q + t$ has no denominator, it suffices to compute the cancellation in the first term of the previous expression.

Property iii) of the Frobenius map is, as noted in its proof, equivalent to the fact that $\mathbb{F}_q$ consists precisely of the q roots of $t^q - t$. Hence

$$t^q - 1 = \prod_{\alpha \in \mathbb{F}_q} (t - \alpha),$$

so to compute d_{-1} we wish to cancel all common factors of the fraction

$$\frac{(t^3 + at + b)[(t^3 + at + b)^{(q-1)/2} + 1]^2}{\prod_{\alpha \in \mathbb{F}_q} (t - \alpha)^2}.$$

The only factors to cancel from the denominator of this quotient are either

i) $(t - \alpha)^2$ with $(\alpha^3 + a\alpha + b)^{(q-1)/2} = -1$ or

ii) $t - \alpha$ with $\alpha^3 + a\alpha + b = 0$.

[Recall that $t^3 + at + b$ has no repeated root.] Let

$$m = \text{ the number of factors of the first kind,}$$
$$n = \text{ the number of factors of the second kind.}$$

Since factors of the first kind are coprime to the factors of the second kind,

$$d_{-1} = 2q + 1 - 2m - n.$$

Since $d_0 = q$, this gives

$$d_{-1} - d_0 - 1 = q - 2m - n. \tag{10.28}$$

Now an α in $\mathbb{F}_q$ with $\alpha^3 + a\alpha + b$ equal to a nonzero square in $\mathbb{F}_q$ will give two solutions of $y^2 = x^3 + ax + b$, whereas there is only one solution of this equation when $\alpha^3 + a\alpha + b = 0$. Moreover, Euler's criterion says that $\alpha^3 + a\alpha + b$ is a nonsquare if and only if $(\alpha^3 + a\alpha + b)^{(q-1)/2} = -1$, so m counts the number of α which do not correspond to any solution of $y^2 = x^3 + ax + b$. Hence

$$N_q = 2q - n - 2m,$$

or

$$N_q - q = q - 2m - n. \tag{10.29}$$

Equation (10.26) follows from (10.28) and (10.29). $\qquad\square$

Theorem 10.14. *The degree function $d(n)$ is a polynomial of degree 2 in n. In fact,*

$$d(n) = n^2 - (d_{-1} - d_0 - 1)n + d_0. \tag{10.30}$$

Proof. By induction on n. For $n = -1$ and 0, (10.30) is a triviality. By the Basic Identity and the induction hypothesis,

$$
\begin{aligned}
d_{n+1} &= 2d_n - d_{n-1} + 2 \\
&= 2[n^2 - (d_{-1} - d_0 - 1)n + d_0] \\
&\quad - [(n-1)^2 - (d_{-1} - d_0 - 1)(n-1) + d_0] + 2 \\
&= (n+1)^2 - (d_{-1} - d_0 - 1)(n+1) + d_0.
\end{aligned}
$$

The induction step in the other direction can be carried out in a similar manner. $\qquad\square$

Proof of the Riemann Hypothesis

We consider the roots x_1, x_2 of the quadratic polynomial

$$d(x) = x^2 - (N_q - q)x + q.$$

Suppose that (10.19) fails to hold, so that the discriminant $(N_q - q)^2 - 4q$ is positive. Then x_1, x_2 are distinct real numbers, say $x_1 < x_2$. By the way it is constructed, $d(x)$ takes only nonnegative integer values on $\mathbb{Z}$, so there must exist some $n \in \mathbb{Z}$ such that

$$n \le x_1 < x_2 \le n + 1. \tag{10.31}$$

Since the coefficients of $d(x)$ are in $\mathbb{Z}$, we have $x_1 + x_2$, $x_1 \cdot x_2 \in \mathbb{Z}$. Hence

$$(x_1 - x_2)^2 = (x_1 + x_2)^2 - 4x_1x_2 \in \mathbb{Z},$$

and for (10.31) to hold, we must have $x_1 = n$, $x_2 = n + 1$. But we note that $x_1 x_2 = q$ is a prime power, so this could only happen if $q = 2$ and $n = 1$ or -2, which is a contradiction since we have assumed throughout that $p \neq 2$. We thus conclude that (10.19) must hold, as desired.   □

11

Epilogue: Fermat's Last Theorem

11.1   Fermat's Last Theorem

This book began with the discussion of Fermat's Last Theorem and it is appropriate to end it with the news that FLT is now a proven theorem, thanks to the work of numerous mathematicians. The strategy for this proof was suggested by G. Frey [17]. Based on some heuristic arguments, he concluded that if FLT is false for an odd prime $\ell > 5$, i.e. if

$$a^\ell + b^\ell = c^\ell$$

holds for integers a, b, c where $abc \neq 0$, then the curve (now called the *Frey curve*)

$$y^2 = x(x - a^\ell)(x + b^\ell)$$

will have some properties which would assure it cannot exist. The so-called Taniyama-Shimura conjecture (1955–1957), which became widely known through a paper of Weil in 1967 asserts that every elliptic curve defined over $\mathbb{Q}$ is modular. In 1986, Ribet [30] proved that the truth of the Taniyama-Shimura conjecture for a certain class of elliptic curves (the semistable case) would be enough to imply FLT. It is this special case of the Taniyama-Shimura conjecture which in 1994 Sir Andrew Wiles proved (cf. [42]) to complete the proof of FLT. However, by the work of Breuil et al. [6], the Taniyama-Shimura conjecture is now a proven fact called the *Modularity Theorem*.

To give a vague idea of the Taniyama-Shimura conjecture, recall [8, Appendix] that an elliptic curve $E/\mathbb{C}$ defined by its Weierstrass equation

$$y^2 = 4x^3 - g_2 x - g_3$$

can be realized as a torus $\mathbb{C}/L$ for a suitable lattice L in $\mathbb{C}$. The map $\phi : \mathbb{C}/L \overset{\cong}{\to} E$ given by $\phi(z) = (\wp(z), \wp'(z))$ parameterizes E by the Weierstrass $\wp$-function and its derivative, which are invariant under the translation by elements of L, i.e. $\wp(z + w) = \wp(z)$ for w in L. Similarly, a modular elliptic curve is an elliptic curve $E/\mathbb{Q}$ that has something similar to do with the action of the modular group $\Gamma_0(N)$ on the upper half plane $\mathfrak{h} = \{z \in \mathbb{C} \mid \mathrm{Im}(z) > 0\}$.

By definition, $\Gamma_0(N)$ is the group of 2×2 matrices

$$M = \begin{pmatrix} a & b \\ c & d \end{pmatrix}$$

over $\mathbb{Z}$, of determinant 1 with c divisible by the integer $N \geq 1$. The aforementioned action of $\Gamma_0(N)$ on $\mathfrak{h}$ is given by the fractional linear transformations

$$z \to Mz = \frac{az + b}{cz + d}.$$

Roughly speaking, a *modular function* of *weight* k and *level* N is an analytic function $f(z)$ on the upper half plane $\mathfrak{h}$ with some growth conditions such that

$$f(Mz) = (cz + d)^k f(z)$$

for all M in $\Gamma_0(N)$.

An elliptic curve $E/\mathbb{Q}$ has an equation

$$y^2 = x^3 + Ax + B \quad (A, B \in \mathbb{Z}) \tag{11.1}$$

with its discriminant $\Delta = -16(4A^3 + 27b^2)$ minimal. Its *conductor* N is a certain product of primes dividing Δ. The *modularity conjecture* states that the points on such an elliptic curve are parameterized by modular functions $f(z)$, $g(z)$ of weight 2 and level N. In other words, (11.1) may be written as

$$(g(z))^2 = (f(z))^3 + A f(z) + B.$$

11.2   An Alternative Approach to Proving FLT

The following conjecture of Masser (1985) and Oesterlé (1988), also called the abc conjecture offers, apparently, a simpler approach for proving FLT for sufficiently large exponents.

Suppose an integer $n > 0$ is a product of powers of distinct primes $p_1, \ldots, p_r$. The *radical* of n is the squarefree integer $\mathrm{rad}(n) = p_1 \ldots p_r$.

The abc conjecture. Suppose for mutually coprime integers a, b, c with $abc \neq 0$ we have $a + b = c$. Given $\epsilon > 0$, there is a constant $k = k(\epsilon)$ such that

$$\max(|a|, |b|, |c|) \leq k(\epsilon)\mathrm{rad}(abc)^{1+\epsilon}.$$

Proof of FLT. If u, v, w is a nontrivial solution of $x^n + y^n = z^n$ with u, v, w mutually coprime, we may assume that $\max(|u|, |v|, |w|) = |w|$. By the abc conjecture,

$$|w|^n \leq k(\epsilon)\mathrm{rad}(|w|)^{3(1+\epsilon)}.$$

This shows that FLT can fail only for a bounded set of exponents n.   □

Bibliography

[1] E. Artin, *Collected Papers*, Addison-Wesley, Reading, MA (1965).

[2] E. Artin, *Galois Theory*, Dover, Mineola, NY (1998).

[3] M. F. Atiyah and I. G. Macdonald, *Introduction to Commutative Algebra*, Addison-Wesley, Reading, MA (1969).

[4] E. Bombieri, Counting points on curves over finite fields (d'apres S.A. Stepanov) *Séminaire Bourbaki*, vol. 1972/1973, exp. 430, in Lecture Notes in Mathematics, vol. 383, Springer, Berlin (1974).

[5] Z. I. Borevich and I. R. Shafarevich, *Number Theory*, Academic Press, London (1966).

[6] C. Breuil, B. Conrad, F. Diamond, and R. Taylor, On the modularity of elliptic curves over $\mathbb{Q}$: Wild 3-adic exercises, *J. Amer. Math. Soc.* **14** (2001), 843–939.

[7] J. W. S. Cassels and A. Fröhlich (Eds.), *Algebraic Number Theory*, Academic Press, London (1967).

[8] J. S. Chahal, *Topics in Number Theory*, Plenum, New York (1988).

[9] J. S. Chahal, *Fundamentals of Linear Algebra*, CRC Press, Boca Raton (2018).

[10] J. S. Chahal, Manin's proof of the Hasse inequality revisited, *Nieuw. Arch. Wiskd*, **13** (1995), 219–232.

[11] R. Dedekind, *Theory of Algebraic Integers*, Cambridge Univ. Press, Cambridge (1996).

[12] P. Deligne, La conjecture de Weil, *Publ. Math. I.H.E.S.* **43** (1974), 273–307.

[13] L. E. Dickson, Fermat's last theorem and the origin and nature of algebraic number theory, *Ann. Math.*, **18** (1917), 161–187.

[14] B. Dwork, On the rationality of the zeta function of an algebraic variety, *Amer. J. Math.*, **82** (1960), 631–648.

[15] J. Esmonde and M. Ram Murty, *Problems in Algebraic Number Theory*, Springer, New York (1999).

[16] E. Freitag and R. Kiehl, *Etale Cohomology and the Weil Conjecture*, Springer, Berlin (1988).

[17] G. Frey, Links between stable elliptic curves and certain diophantine equations, *Annales Universitatis Saraveinsis*, **1** (1986), 1–40.

[18] C. F. Gauss, *Disquisitiones Arithmeticae*, Springer, Berlin (1985).

[19] M.J. Greenberg, An elementary proof of the Kronecker-Weber theorem, *American Math. Monthly*, **81** (1974), 601–607.

[20] A. Grothendicck and J. Dieudonné, *Éléments de géométrie algébrique*, Pub. Math. I.H.E.S. (1960–1967).

[21] H. Hasse, *Vorlesungen über Klassenkörpertheorie*, Physica-Verlag, Würzburg (1967).

[22] E. Hecke, *Lectures on the Theory of Algebraic Number Fields*, Springer, New York (1981).

[23] I. N. Herstein, *Topics in Algebra*, Blaisdell, New York (1964).

[24] D. Hilbert, *The Theory of Algebraic Number Fields*, Springer, Berlin (1998).

[25] K. Ireland and M. Rosen, *A Classical Introduction to Number Theory*, Springer, New York (1990).

[26] Yu. I. Manin, On cubic congruences to a prime modulus, *Izv. Akad. Nauk USSR, Math. Ser.* **20** (1956), 673–678.

[27] D. A. Marcus, *Number Fields*, Springer, New York (2018).

[28] J. Neukirch, *Class Field Theory*, Springer, Berlin (1986).

[29] T. Ono, *An Introduction to Algebraic Number Theory*, Plenum, New York (1990).

[30] K. Ribet, From the Taniyama-Shimura Conjecture to Fermat's Last Theorem, *Anales de la faculté des Sciences de Toulouse*, **11** (1990), 116–139.

[31] J. Rotman, *Galois Theory*, Springer, New York (1998).

[32] K. Rubin and A. Silverberg, A Report on Wile's Cambridge Lecture, *Bull. (New Series) Amer. Math. Soc.*, **31** (1994), 15–38.

[33] P. Samuel, *Algebraic Theory of Numbers*, Hermann, Paris (1970).

[34] F. K. Schmidt, Analytische Zahlentheorie in Körpern der Charakteristik p, *Math Z*, **33** (1931), 1–32.

[35] W. M. Schmidt, Zur Methode von Stepanov, *Acta Arith.*, **24** (1973), 347–367.

[36] W. M. Schmidt, *Equations over Finite Fields: An Elementary Approach*, Kendrick Press, Heber City, UT (2004).

[37] L. Washington, *Elliptic Curves: Number Theory and Cryptography*, CRC Press, Boca Raton (2008).

[38] A. Weil, Number of solutions of equations in finite fields, *Bull. Amer. Math. Soc.*, **55** (1952), 497–508.

[39] A. Weil, *Basic Number Theory*, Springer, New York (1967).

[40] A. Weil, *Number Theory for Beginners*, Springer, New York (1979).

[41] A. Weil, *Number Theory: An Approach through History*, Birkhäuser, Boston (1984).

[42] A. Wiles, Modular elliptic curves and Fermat's Last Theorem, *Ann. Math.*, **141** (1995), 443–551.

Index

Printed in the United States
by Baker & Taylor Publisher Services